T0320006

Social Dialogue in the Gig Economy

Social Dialogue in the Gig Economy
A Comparative Empirical Analysis

Edited by

Jean-Michel Bonvin

Professor of Sociology and Social Policy, Department of Sociology, University of Geneva, Switzerland

Nicola Cianferoni

Scientific Officer, State Secretariat for Economic Affairs SECO, and Senior Researcher, University of Geneva, Switzerland

Maria Mexi

Senior Researcher and Fellow, Albert Hirschman Centre on Democracy, Chair, Digital Innovation – Global Excellence Network/Tech Hub, Graduate Institute of International and Development Studies, Switzerland, and consultant, International Labour Organization

Edward Elgar
PUBLISHING

Cheltenham, UK • Northampton, MA, USA

Published by
Edward Elgar Publishing Limited
The Lypiatts
15 Lansdown Road
Cheltenham
Glos GL50 2JA
UK

Edward Elgar Publishing, Inc.
William Pratt House
9 Dewey Court
Northampton
Massachusetts 01060
USA

A catalogue record for this book
is available from the British Library

Library of Congress Control Number: 2022948746

This book is available electronically in the **Elgar**online
Sociology, Social Policy and Education subject collection
http://dx.doi.org/10.4337/9781800372375

ISBN 978 1 80037 236 8 (cased)
ISBN 978 1 80037 237 5 (eBook)

Printed and bound by CPI Group (UK) Ltd, Croydon, CR0 4YY

Contents

Contributors

Simone Baglioni holds a chair in sociology at the University of Parma. His research interests focus on migration and employment, precarious work and unemployment, civil society and collective action, solidarity, and social innovation. He is the coordinator and principal investigator of the Horizon 2020 project Sirius (www.sirius-project.eu).

Jean-Michel Bonvin is Professor of Sociology and Social Policy at the University of Geneva. His main fields of expertise include social policies, sociology of work, and economic sociology. His research has been funded, among others, by the Swiss National Science Foundation (SNSF) and by the European Commission. He has extensively published in leading international journals such as *European Societies*, *Social Policy and Society*, *Review of Social Economy*, *Policy & Politics*. He co-coordinated the Swiss Network for International Studies (SNIS)-funded project Gig Economy and Its Implications for Social Dialogue and Workers' Protection.

Nicola Cianferoni is a sociologist of work and holds a PhD in socio-economics. He works as a scientific collaborator in the Work and Health section of the Swiss State Secretariat for Economic Affairs SECO and is associate researcher at the Institute for Sociological Research (IRS) of the University of Geneva.

Johannes Kiess, PhD, is a PostDoc at the Department of Social Sciences of the University of Siegen and Deputy Director of the Else Frenkel-Brunswik Institute for Democracy Research at Leipzig University. His research interests and teaching subjects include industrial relations, trade unionism, European integration, political mobilization, and political attitudes with a special focus on right-wing extremism.

Maria Mexi is a senior researcher and fellow at the Albert Hirschman Centre on Democracy and the Chair of the Digital Innovation – Global Excellence Network/Tech Hub at the Graduate Institute of International and Development Studies, Switzerland, and a consultant at the International Labour Organization. She formerly worked at the University of Geneva, where she co-coordinated and conducted research for the SNIS-funded project Gig Economy and its Implications for Social Dialogue and Workers' Protection, as well as several Horizon 2020 international projects. Dr. Mexi has worked as a government advisor in several countries around the world, as well as an expert for the European Commission.

Tom Montgomery is a research fellow at the Yunus Centre for Social Business and Health and a lecturer in politics at Glasgow Caledonian University. His research focuses on issues of solidarity, civil society, youth employment, and the future of work. His current research focuses on issues of labour market integration and the gig economy.

Konstantinos Papadakis is Senior Specialist at the Governance & Tripartism Department of the ILO. From 2013 to 2016, he was Senior Liaison Officer for Cyprus and Greece and in charge of ILO Athens. Previously, he was Adviser to the Executive Director of the ILO's Social Dialogue sector (2012–2013); Research and Policy Development Specialist at the Industrial Relations Department (2008–2013); and Researcher at the International Institute of Labour Studies (2000–2008). He has published extensively on national and cross-border industrial relations, and has been the main author of the International Labour Conference reports on social dialogue (2013 and 2018) and of the report debated at the Meeting of Experts on cross-border social dialogue (2019). He holds a PhD in international relations (Higher Educational Institute, IUHEI, Geneva, 2006).

Luca Perrig is an economic sociologist and PostDoc researcher at the University of Geneva. In his PhD, he studied online food deliveries in Western Switzerland, with a focus on the digitization of labour markets.

Acknowledgements

The idea of this book originated in the research project Gig Economy and its Implications for Social Dialogue and Workers' Protection, funded by the Swiss Network for International Studies (SNIS). This project provided an ideal environment for the development of our ideas and reflections. The editors and chapter authors express their deepest gratitude to the SNIS for providing such an opportunity and for its support during the project. The information and views set out in this volume are those of the authors and do not necessarily reflect the official opinion of the SNIS.

We also owe warm thanks to Sven Carlström, whose editorial support in the preparation of the book and in thoroughly checking the references proved essential. Last but not least, we are also grateful to Finn Halligan and Alex Pettifer of Edward Elgar Publishing, who followed our work with great benevolence and encouragement throughout the process.

1. Introduction

Jean-Michel Bonvin, Nicola Cianferoni and Maria Mexi

Developments in the *gig economy* have attracted immense attention in academic and policymaking circles recently. The gig economy consists both of work that is transacted via platforms but delivered in a specific locality and of platforms that enable remote working (Wood et al. 2019a; Drahokoupil 2021). These platforms, operating either in the *global* or in the *local* gig economy (see Mexi and Papadakis' chapter in this book), act as market intermediaries that significantly reduce the overhead costs of outsourcing microtasks ('gigs') and offshoring by providing a standardized framework for identifying, contracting and paying workers. It is widely considered that the Covid-19 pandemic has expanded the use of digital platforms, giving rise to innovative ways of remote working and offering flexibility for both workers and businesses, but also exacerbating precarious working conditions (Mexi 2020a). As the gig economy is expanding in size and importance, uncertainty around the social outcomes remains while observers regularly fall prey to bouts of either optimistic or pessimistic views on its disruptive impact. Sceptics claim that, while many workers are enjoying increased income and autonomy, the manner in which gig economy platforms operate often results in low pay, social isolation, working unsocial and irregular hours, overwork, sleep deprivation and exhaustion (Wood et al. 2019a). While these issues are not new in the world of work, gig workers face some unique barriers when it comes to rectifying them.

During rapid economic and labour market changes in the past, workers formed labour organizations to advocate for their needs and interests. But that organizing often presumed "a single employer, a single workplace, and a set of duties and obligations that can be structured around a contract that stays in place for several years" (Sabeel Rahman 2016, p. 11). In the gig economy, given the moves away from full-time work and direct employment, these factors seem to have lost significance. It is also the case that national labour laws are not applied, or only to some extent, to gig workers in several countries. This issue is particularly acute in the *global* gig economy, where platforms' operations and functioning resemble those of multinational companies, involving transactions across national borders, and it becomes unclear which jurisdictional regulations apply. Overall, pessimistic and Polanyi-inspired accounts of the gig economy stress that, due to its *dis-embedding* from institutional interventions (particularly welfare states and strong trade unions), labour in the gig economy is increasingly being *re-commodified*, a process that intensifies the disciplinary power of labour market competition. As Wood et al. (2019b, p. 936) suggest, "the pendulum has swung from commodification to de-commodification and back again."

Concerns about these negative outcomes have led to both calls for policy interventions and actual interventions by the social partners in several countries. These interventions take a variety of shapes and contents in order to promote what could be called "decent digiwork" (Mexi 2020b). They are faced with difficulties induced by the specific context of the gig economy, especially with regard to overcoming the manifold risks of fragmentation that the organization of collective action faces in that context (Heiland 2020). Against this background, this introduction is articulated in two main sections. The first one enumerates the key labour issues raised by the gig economy as they are framed in the international policy debate and the academic literature. The second focuses on the main obstacles faced by social dialogue within the gig economy and how emerging initiatives try to tackle them (Lenaerts et al. 2018; Joyce and Stuart 2021; Vaughan-Whitehead et al. 2021). We conclude with a brief presentation of the overall objectives of the volume and the content of the book chapters.

1.   WHAT IS AT STAKE WITH THE GIG ECONOMY?

The digital transformation of our economies is evolving at an exponential pace, fundamentally altering the way we work and live. Across the globe, advances in information and communication technology have redefined labour markets as supply chains are unbundled into transnational networks of buyers and suppliers. The digitization of the global economy is seen particularly in the rise of the gig economy: digitally enabled work arrangements that defy many long-held understandings of employment relationships and labour protections in the formal sector. More specifically, long-term employment relationships tend to be replaced by one-off transactions bearing on tasks, thus transforming "labour markets from markets for jobs to markets for tasks" (Drahokoupil and Vandaele 2021, p. 1) and implying the loss of protections and rights related to long-term employment relationships.

The gig economy has grown rapidly over the past decade. Based on data presented in the ILO's *World Employment and Social Outlook* report (2021), globally, platforms – both online web-based (involving micro-tasks, freelance and competitive programming) and location-based such as in the taxi and delivery sectors – rose from 142 in 2010 to over 777 in 2021. Thus, gig economy transactions are increasingly playing a critical role in reshaping the global economy (Charlton 2021). It is widely acknowledged that the impact of the gig economy extends much beyond its contribution as a supplemental income for its workforce, as it induces new forms of organizing work and value chains in a global context. In this line, the European Political Strategy Centre states that 'work' is being redefined in unprecedented ways: it is no longer a static concept but an umbrella term for tasks and roles performed in a different manner and under various and different legal arrangements (European Political Strategy Centre 2016). Such developments have a potential impact on the whole world of work (Valenduc and Vendramin 2016, p. 33).

In parallel to a positive understanding of the effects of the gig economy and work drawing primarily on the 'freedom' and 'flexibility' discourse, a growing body of literature (e.g. Aloisi 2015; Graham and Shaw 2017; Huws 2017; Schmidt 2017) is devoted to analyzing the many unresolved issues arising from the gig economy, including questions of commodification and outsourcing of work tasks, casualization/informalization of employment along temporal, spatial and institutional dimensions, ICT-algorithm-enabled monitoring and management of work, and liberalization of the regulation of work. From this perspective, the demand for paying attention to the gig economy comes from the recognition that gig work amplifies longer-standing trends in the informal economy, including critical social protections gaps.

A large number of existing studies in the literature on the gig economy have primarily focused on labour issues. Several authors have argued that work in the gig economy has similarities with very old working arrangements, such as *putting-out work*/*homework*[1] and *piecework*[2] (Prassl and Risak 2016; De Stefano 2016; Stanford 2017) and that it also closely resembles other types of flexible and non-standard employment such as temporary work, part-time work or temporary agency work (ILO

2016). Research has provided evidence for the growth in insecure and precarious work and non-standard forms of employment, and how this potentially hinders long-term incentives for innovation (Kalleberg 2000, 2011; Standing 2015; Malhotra and Van Alstyne 2014), long before the rise of the digital, platform-based, gig economy. Hence, a distinct literature on labour market precariousness in the gig economy seems to have evolved in parallel with – and inspired from – the relevant literature in the traditional economy (see, for example, Smith and Leberstein 2015; JRC 2016).

Overall, in the surveyed literature there are concerns that many social protections (unemployment protection, pension rights or maternity leave) within the EU are not accessible by 'atypical workers' but only by those who meet a legal definition of 'employee'. Gig jobs have often engulfed employment relationships that do not take health and safety or social security considerations into account, leaving workers without basic social protection with serious implications to their economic and social security. This view is aligned with the experiences of workers classified under the term 'cybertariat' (Huws and Sturman 2004). In a similar vein, Codagnone et al. (2016) argue that the digital platforms create contingent and precarious employment forms. De Stefano (2016, p. 8) points out that, even in those cases where gig workers are classified as employees, "the intermittent nature of their activity could be an obstacle to accede to important employment or social rights [...] when these rights are dependent upon a minimum length of service." MacDonald and Giazitzoglu (2019) have empirically demonstrated how young adults' experiences with the gig economy in the United Kingdom are typified by a lack of control and choice, disempowerment, insecurity, low pay, and degraded work conditions. Very often precarious gig jobs become 'traps' instead of 'bridges' into secure work, thereby reducing social mobility (ILO 2020). In addition, some platforms have significant gender pay gaps (Liang et al. 2018). Piasna et al. (2022) have emphasized the income precariousness of platform workers in the EU (a mean of 250 euros per month across 14 EU Member States), with only 7.5% of remote workers and 7% of on-location workers relying almost completely on platform for their earnings.

Hence, the precariousness and insecurity of platform workers is documented by a number of empirical studies. This suggests a need for adapting existing labour and social protection systems to the specific conditions of such workers, so as to realize the human right to decent work and social security for all. Most pressing areas of concern are threefold:

(a) The status of gig workers: Are they wage earners, thus enjoying the full benefits of labour law and social protection regimes? Or are they self-employed workers who have to bear the entrepreneurial risk connected to their economic activity, and are not entitled to social benefits or labour protection? Or should a specific interme- diary status be innovated for them, such as 'dependent contractor' (Taylor 2017) or 'independent worker' (Aloisi 2015; Harris and Krueger 2015)? This has significant implications: if a person working in the digital, platform-based, gig economy is considered a freelancer, self-employed, and the platform company itself more an *intermediary* than an *employer stricto sensu*, then labour law becomes inapplicable. In many instances, this has resulted in situations of worker misclassification that have led to workers being denied statutory protection such as minimum wage, health and safety protections, or paid annual leave. Within this debate, a central point of contention concerns the degree of subordination or autonomy over working time.

(b) Access to social protection: To secure gig workers' entitlement to social rights, proposals to decouple access to social protection benefits from full employment are being formulated, such as lowering the minimum period of employment required to qualify for employment-related protection, "allowing more flexibility with regard to interrupted contribution periods" or ensuring "portability of entitlements" for a flexible workforce (ILO 2016, p. 302).

(c) Enforcement of existing labour standards: This is hindered by the fact that gig workers work in isolated settings and generally have limited knowledge about their rights (Stewart and Stanford 2017; Degryse 2016, 2017). In this regard, better enforcement of labour standards entails not only enhanced regulatory control, but also raising awareness and educating workers about their rights and reinforcing collective representation mechanisms.

The purpose of our book is to investigate the potential contribution of social dialogue mechanisms to tackle such issues and promote genuine access to decent digiwork and social protection to all gig workers. The next section identifies the obstacles that the implementation of social dialogue faces in the specific context of the gig economy.

2.   SOCIAL DIALOGUE AND THE GIG ECONOMY: OBSTACLES AND OPPORTUNITIES

Issues of gig workers' precariousness and limited access to labour and social protection have so far monopolized the academic and policy debate. At the same time, a new repertoire of contention is emerging. Social partners – who have valuable, sector-specific knowledge that is vital to level the playing field – are increasingly involved and mobilized on these issues (Tassinari and Maccarrone 2020), bringing pressure for more fine-tuned regulation and pushing digital platforms to come to the negotiation table. This in turn makes it critical that gig workers are enabled to have a voice at work and in policy debates around the future of the world of work. Indeed, for gig workers in unbalanced power relationships with digital platforms (Mexi 2020b), worker organizing, the development of agency, voice and representation and its expression through collective bargaining are a promising route to a more inclusive future (Johnston and Land-Kazlauskas 2018). Beyond the academic literature, this has been confirmed by the work of the Organisation for Economic Co-operation and Development (OECD 2020) and the International Labour Organization (ILO/AICESIS/OKE Report 2018), as well as by the European Commission's discussions with social partners on how to regulate platform work (see for instance the proposal for an EU directive issued in 2021 – EC 2021). Empirical evidence (OECD 2019) shows that coordinated bargaining systems are linked with less wage inequality and higher employment. Whether considering issues of workplace adjustment to new technologies or job quality, workers' representation and collective bargaining arrangements constitute key tools enabling governments and social partners to find and agree on fair, tailored solutions.

Despite this wide-ranging consensus on the importance and potential of social dialogue to regulate the gig economy, the ways in which social dialogue can contribute to a more inclusive digital landscape in the growing gig economy are less explored in academic studies. The purpose of our book is to fill this gap and help shape momentum to set in place national or global frameworks (Albrecht et al. 2021) for structured dialogue and collective bargaining among governments, platform businesses and workers, as part of a broader strategy to democratize the platform economy as a whole – from its governance to the ability of individual workers to organize and make decisions together about their work. This calls for making a precise diagnosis of the present situation with regard to social dialogue and the opportunities and obstacles it faces within the gig economy.

In the past, workers were able to form organizations to advocate in support of their interests and respond to challenges raised by economic developments. Today, when it comes to building collective agency, gig workers face unique barriers. Mobilizing and organizing collectively when work is digital, discontinuous and globally dispersed poses challenges. In the gig economy, given the moves away from full-time work and direct relationship with an identifiable employer, conditions for collective organization are more difficult to meet. Another important factor complicating the capacity to organize is the disparity of work performed by different segments of the workforce across various platforms. Workers are often tied to a multitude of platforms (Prassl 2018), which translates into starkly heterogeneous worker motivations, experiences and claims, constraining the leverage offered by effective collective action and representation of interests. In this context, studies drawing on the *insider–outsider* theory (Lindbeck and Snower 2002; Rueda 2005) have brought information on how established workers (*insiders*) have used their bargaining power to protect their employment. Conversely, *outsiders*, such as young and unemployed or gig workers, have lacked political influence to leverage change, either at their immediate workplace or beyond. Critics argue that mainstream trade unions almost exclusively represent permanent and full-time workers (Standing 2015). Pursuing this line of reasoning, Heiland (2020) emphasizes that collective organization and action in the gig economy is faced with five kinds of fragmentation:

(a) A legal fragmentation resulting from the self-employed status, which is commonly used for gig workers and impedes both collective organization, considered as a breach to antitrust laws, and access to social and labour protection rights that are reserved to those with wage earner status. Another source of legal fragmentation derives from the national boundaries of labour legislation, which allow platforms choosing the most favourable location for their activities or locating them in the digital space, beyond any national jurisdiction, thus implementing a kind of 'regime shopping' (Zwick 2018).

(b) A spatial fragmentation that touches more specifically crowd digital work, as geographically dispersed workers on platforms such as Amazon Mechanical Turk have difficulties in collectively organizing and enacting solidarity. This issue applies, though to a lesser extent, to on-location gig workers, who are also left without formal meeting places or communication channels and have to build their own alternative ways to meet and communicate and represent a decentralized workforce (Soderqvist 2017).

(c)   An organizational fragmentation whereby cooperation between workers is not necessary in the gig economy. Rather, competition between gig workers is promoted, as they are separated from each other in the labour process (just like in the Taylorian firm, but in a way that is further enhanced by spatial fragmentation and the fact that they do not work in the same factory) and they have to compete for each gig. This creates an environment where collective organization is not encouraged. Organizational fragmentation is further reinforced by the fact that platforms claim to act as matchmakers between gig workers and consumers; as a consequence, conflicts have to be negotiated in a more complex configuration involving all these stakeholders.

(d)   A technological fragmentation related to algorithmic management that minimizes the necessity of communication and makes human management almost invisible, thus undermining possibilities of resistance.

(e)   A social fragmentation emphasizing the social heterogeneity of gig workers, making it difficult to build coalitions and collective organizations between such a diversity of interests. This may reinforce the dividing line between insiders and outsiders mentioned above and illustrates how such conflicting interests may also be present within the gig economy itself and not only between gig workers and 'traditional' workers.

Despite such obstacles, fresh approaches are being developed by trade unions or gig workers themselves (Countouris and De Stefano 2019) to adapt to the changing conditions of platform work: opening union membership to platform workers, establishing new initiatives and unions, and/or negotiating collective agreements with platform companies. Most common initiatives take one of the following three forms:

–   The first is inspired by a *legal approach*, which entails unions contesting worker misclassification with a view to including gig workers in existing employment arrangements and social entitlements. The objective is to rely on court decisions to promote the recognition of the wage earner status for gig workers, thus securing their access to labour and social rights. Examples of successful court litigation strategies include the efforts led by GMB – one of the largest trade unions in the United Kingdom – which resulted in over 30,000 drivers being granted access to basic employment benefits, including minimum wage and paid leave (McGoogan and Yeomans 2016, see also Montgomery and Baglioni in this book). Litigation is considered one crucial strategy to empower precarious workers and to show that there is no real distinction between those working via platforms and those who are not (Roberts 2018 – see also Bonvin, Cianferoni and Perrig in this volume showing how recent court decisions in the Swiss context converge to consider gig workers as wage earners).

– The second involves *alliance formation* – that is, creating collective organizations defending gig workers' labour and social rights. This takes place either via (self-)mobilization of gig workers themselves or via their representation by existing trade unions. Such initiatives relate to two different ways of conceiving collective organization and representation: the logic of membership on the one hand where the objective is worker organization; the logic of influence on the other hand where efforts are rooted in the institutional logic of the national industrial relations systems (Vandaele 2021). Hence, grass-roots initiatives, often with horizontal and combative features, coexist with conventional trade union actions, characterized by more vertical and bargaining-oriented operational modalities (Lenaerts et al. 2018). Joyce and Stuart (2021) illustrate how both types of initiatives may be disconnected, leading to problems of coordination. The intensity of such problems varies according to the national configuration of industrial relations as is evidenced by the Leeds Index of Platform Labour Protest (Trappmann et al. 2020).

– The third approach encompasses efforts for *advocacy and regulatory reform* (through lobbying, campaigning and influencing) aimed at putting pressure for introducing new legislation at national and local levels. Successful examples at local level include the Teamsters in Seattle, who were able to pressure for new legislation extending collective bargaining to include gig workers in the transport sector (most notably for Lyft and Uber drivers) (Kessler 2017). This strategy also takes place at international level, as illustrated by the EU-level social dialogue that involved social partners from individual sectors and resulted in the conclusion of a European framework agreement on digitalization in March 2020 (Vaughan-Whitehead et al. 2021). This raises the issue of how actions taken at local or national level, embedded in specific national industrial relations configurations, connect with and translate into international initiatives extending beyond national boundaries, possibly also in the virtual space (see Mexi and Papadakis in this volume).

Hence, the implementation of social dialogue is faced with a variety of difficulties related to the risks of fragmentation and the lack of coordination between the diverse initiatives taking place at grass-roots or institutional level and at local, national or global level. The focus of our book is to address these issues empirically and to see whether and to what extent social dialogue takes place despite these difficulties and the variety of forms it takes in different countries in order to overcome these obstacles.

3.   PURPOSE AND OUTLINE OF THE BOOK

The initial idea of the book originates in the project Gig Economy and its Implications for Social Dialogue and Workers' Protection, funded by the Swiss Network for International Studies (SNIS).[3] This volume offers different perspectives on how the gig economy is reshaping the world of work national and globally. More specifically, it asks: What are the experiences of individuals undertaking gig economy work and the challenges they face? How can public policy actors and social partner organizations transform and adapt their interventions in response to the opportunities and risks of gig work in different national contexts and sectors? How can governments and international agencies put in place effective frameworks that can lead to rebalancing power asymmetries in platforms' cross-border operations, and support more inclusive social protection models? Scholars and international researchers and practitioners who have contributed to this volume analyze the type of gig jobs and work, the role and responses of governments and social partners at national and sectoral level, and the specific social dialogue dynamics that take place in various national industrial relations configurations. This will, in turn, support a more informative and comprehensive framework for better assessing how public policy actors and social partners can manage and respond to the impact of gig work on the labour market.

    In Chapter 1, Johannes Kiess shows how the coordinated German model of capitalism copes with the disruptive force of the gig economy. Based on extensive interview material, the chapter emphasizes that the gig economy's impact on the German model is limited by existing institutions and regulations, and adaptations to the challenges of digitalization remain incremental and path-dependent to a large extent. In Chapter 2, Maria Mexi focuses on social partners' (non-)responses to the challenges posed by the gig economy in the Greek context. Her empirical findings indicate the importance of policy legacies and institutional frames for understanding why the gig economy did not constitute a topic of contention between Greek workers and employers until now and why it has not given rise to a new framework for facilitating social dialogue between the parties involved. In Chapter 3, Jean-Michel Bonvin, Nicola Cianferoni and Luca Perrig explain that the legal and political debates related to the gig economy in Switzerland do not call into question the existing model of industrial relations based on consensual social dialogue and subsidiarity. Nonetheless, their empirical data show that the implementation of social dialogue is facing difficulties when dealing with the gig economy. Innovative strategies and stronger support from public authorities may be needed to overcome these difficulties. In Chapter 4, Tom Montgomery and Simone Baglioni offer a multidimensional analysis on the challenges of the gig economy in the United Kingdom, in that the perspectives adopted cut across macro (policymaking), meso (trade unions and labour organizations) and micro (self-employed workers) levels. They show that standard forms of employment have come under pressure in recent years and how the gig economy reinforces this trend. They underline that debates over the future of work between the unions and the UK government are much more conflictual than in other countries, mostly as a consequence of the austerity policies following the financial crisis. In Chapter 5, Maria Mexi and Konstantinos Papadakis argue that the challenges of the gig economy and the diversity of national experiences, as illustrated in the previous chapters, advocate for a strong regulation of labour platforms at international level. The authors present voluntary cross-border social dialogue initiatives and agreements as a promising path to promote the adoption and implementation of effective labour standards in the gig economy.

The four national cases of Germany, Greece, Switzerland and the United Kingdom presented in this book display a varied spectrum of regulatory, policy and social partnership models, corresponding to different types of European welfare regimes (Esping-Andersen 1990). In particular, Germany falls under the category of 'social partnership' countries corresponding to the 'conservative model' of welfare regimes and displaying a balanced relation of power between workers' and employers' organizations, an integrating bargaining style, and an institutionalized role of social partners in public policy. The United Kingdom belongs to the 'liberal pluralism' group of countries, corresponding with the 'liberal model' of welfare regimes and showing an employer-oriented balance of power between workers' and employers' organizations and company-level bargaining style, and a rare/event-driven role of social partners in public policy. In terms of industrial relations, the UK can be qualified as an example of market-oriented governance (Eurofound 2020). Switzerland's decentralized and consensual industrial relations (Oesch 2010) blurs elements of the German and the UK models (as illustrated by the focus on social partnership and subsidiarity on the one hand and the strong position of employers' unions on the other hand) and its welfare state is often presented as a hybrid case combining features of the liberal, conservative and, to a more limited extent, social-democratic models (Armingeon et al. 2004); while Greece is part of the 'polarized/ state-centered' category of countries, corresponding with the 'Southern European model' of welfare regimes (Ferrera 1996) and displaying an alternating balance of power between workers' and employers' organizations and an irregular/politicized role of social partners in public policy.

Given their different structural configurations, the four countries have responded and adapted differently to the changes in the labour market brought about by the gig economy. Thus, different outcomes on the development and the reform of their social protection and social dialogue institutions and practices are to be expected. All in all, the four country case studies allow understanding and analyzing the key challenges that characterize the varied European landscape of the gig economy – particularly in sectors such as hospitality, transportation and food delivery – both in terms of gig workers' capacity to organize and build collective voice and the degree and level of social partner involvement. The fact remains that the gig economy is an international phenomenon that relies on platform-based business models and has a similar impact on populations and societies. Thus, an international regulation of this phenomenon is needed. The United Nations system – and in particular the International Labour Organization (ILO) – can play an important role in implementing such regulation and in promoting social dialogue as argued in Chapter 5. Both bottom-up and top-down approaches and experiences documented in this book should be considered as complementary. All of them tackle the need to find a way to frame the gig economy and preserve the quality of work in society considering the new realities posed by the digitalization of labour and platform work in the 21st century.

NOTES

1. Referring to a type of subcontracting used at early stages of industrialization whereby work was contracted by a central agent to subcontractors who completed work either in their own homes or in workshops (Risak and Warter 2015).
2. Referring to a form of employment common under the guild system before the 18th century as well as in the industrial era in which a worker was paid a fixed price for each action performed or unit produced irrespective of time.
3. See https://snis.ch/projects/gig-economy-and-its-implications-for-social -dialogue-and-workers-protection/ for more detail on the project. We are grateful to the SNIS for its financial support. Of course all usual disclaimers apply.

REFERENCES

Albrecht, T., K. Papadakis and M. Mexi (2021), "An International Governance System for Digital Labour Platforms", in *The Transformation of Work*, e-book published by Social Europe and Friedrich Ebert Stiftung, pp. 39–44. Available at: https://socialeurope.eu/book/the-transformation-of-work (accessed on 26 September 2022).

Aloisi, A. (2015), "Commoditized Workers. The Rising of On-Demand Work: A Case Study Research on a Set of Online Platforms and Apps", *SSRN Electronic Journal*, **37** (3), 653–90.

Armingeon, K., F. Bertozzi and G. Bonoli (2004), "Swiss Worlds of Welfare", *West European Politics*, **27** (1), 20–44.

Charlton, E. (2021), "What is the Gig Economy and What's the Deal for Gig Workers?", World Economic Forum. Available at: https://www.weforum .org/agenda/2021/05/what-gig-economy-workers/ (accessed on 26 September 2022).

Codagnone, C., F. Biagi and F. Abadie (2016), "The Passions and the Interests: Unpacking the 'Sharing Economy'", JRC Science for Policy Report, Institute for Prospective Technological Studies.

Countouris, N. and V. De Stefano (2019), *New Trade Union Strategies for New Forms of Employment*, Brussels: ETUC.

De Stefano, V. (2016), "The Rise of the 'Just-in-Time Workforce': On-Demand Work, Crowdwork and Labour Protection in the 'Gig-Economy'", Conditions of Work and Employment Series No. 71, Geneva: ILO.

Degryse, C. (2016), "Digitalisation of the Economy and Its Impact on Labour Markets", ETUI Working Paper, ETUI, Brussels.

Degryse, C. (2017), "Shaping the World of Work in the Digital Economy", ETUI Foresight Briefs, January.

Drahokoupil, J. (2021), "The Business Models of Labour Platforms: Creating an Uncertain Future", in J. Drahokoupil and K. Vandaele (eds), *A Modern Guide to Labour and the Platform Economy*, Northampton, MA, USA and

Cheltenham, UK: Edward Elgar Publishing, pp. 33–48. https://doi.org/10.4337/9781788975100.00011.

Drahokoupil, J. and K. Vandaele (2021), "Introduction: Janus Meets Proteus in the Platform Economy", in J. Drahokoupil and K. Vandaele (eds), *A Modern Guide to Labour and the Platform Economy*, Northampton, MA, USA and Cheltenham, UK: Edward Elgar Publishing, pp. 1–31.

Esping-Andersen, G. (1990), *The Three Worlds of Welfare Capitalism*, Princeton, NJ: Princeton University Press.

Eurofound (2020), "Industrial Relations: Developments 2015–2019", Challenges and Prospects in the EU Series, Publications Office of the European Union, Luxembourg.

European Commission (2021), "Proposal for a Directive of the European Parliament and of the Council on Improving Working Conditions in Platform Work", COM(2021) 762 final.

European Political Strategy Centre (2016), "The Future of Work: Skills and Resilience for a World of Change", Issue 13, 10 June.

Ferrera, M. (1996), "The 'Southern' Model of Welfare State in Social Europe", *Journal of European Social Policy*, 6 (1), 17–37.

Graham, M. and J. Shaw (2017), "Towards Another World of Gig Work", in M. Graham and J. Shaw (eds), *Towards a Fairer Gig Economy*, [Online]: Meatspace Press, pp. 4–6.

Harris, S. D. and A. B. Krueger (2015), "A Proposal for Modernizing Labor Laws for Twenty-First-Century Work", Discussion Paper No. 2015-10, Hamilton Project, Brookings, December.

Heiland, H. (2020), "Workers' Voice in Platform Labour", Study No. 21, Hans-Böckler-Stiftung, July.

Huws, U. (2017), "Where Did Online Platforms Come From? The Virtualization of Work Organization and the New Policy Challenges It Raises", in P. Meil and V. Kirov (eds), *Policy Implications of Virtual Work*, Basingstoke, UK: Palgrave Macmillan, pp. 29–48.

Huws, U. and S. Sturman (2004), "The Making of a Cybertariat: Virtual Work in a Real World", *Resources for Feminist Research*, 31 (1–2), 25–33.

ILO/AICESIS/OKE Report (2018), "Social Dialogue and the Future of Work", Report of the ILO–AICESIS Conference, 23–24 November 2017, Athens, Greece.

International Labour Office (2016), "Non-standard Employment Around the World: Understanding Challenges, Shaping Prospects", Geneva: ILO.

International Labour Office (2020), "World Employment and Social Outlook: Trends 2020", Geneva: ILO.

International Labour Office (2021), "World Employment and Social Outlook: The Role of Digital Labour Platforms in Transforming the World of Work", Geneva: ILO.

Johnston, H. and C. Land-Kazlauskas (2018), "Organizing on Demand: Representation, Voice, and Collective Bargaining in the Gig Economy", Working Paper, ILO, Geneva.

Joyce, S. and M. Stuart (2021), "Trade Union Responses to Platform Work: An Evolving Tension between Mainstream and Grassroots Approaches", in

J. Drahokoupil and K. Vandaele (eds), *A Modern Guide to Labour and the Platform Economy*, Northampton, MA, USA and Cheltenham, UK: Edward Elgar Publishing, pp. 177–192.

JRC (2016), "The Future of Work in the 'Sharing Economy': Market Efficiency and Equitable Opportunities or Unfair Precarisation?", JRC Science for Policy Report.

Kalleberg, A. L. (2000), "Non-Standard Employment Relations: Part-Time, Temporary and Contract Work", *Annual Review of Sociology*, **26** (1), 341–365.

Kalleberg, A. L. (2011), *Good Jobs, Bad Jobs: The Rise of Polarized and Precarious Employment Systems in the United States, 1970s to 2000s*, New York: Russell Sage Foundation.

Kessler, S. (2017), "Uber has Produced 18 Episodes of a Podcast Warning Drivers About the Dangers of Joining a Union", *Quartz*, March. Available at: https://qz.com/927777/the-teamsters-have-finally-begun-to-organize-uber-drivers-in-seattle/ (accessed on 26 September 2022).

Lenaerts, K., Z. Kilhoffer and M. Akgüç (2018), "Traditional and New Forms of Organisation and Representation in the Platform Economy", *Work, Organisation, Labour & Globalisation*, **12** (2), 60–78.

Liang, C., Y. Hong, B. Gu and J. Peng (2018), "Gender Wage Gap in Online Gig Economy and Gender Differences in Job Preferences", NET Institute Working Paper No. 18-03.

Lindbeck, A. and D. Snower (2002), "The Insider–Outsider Theory: A Survey", IZA Discussion Paper, No 534, IZA, Bonn.

MacDonald, R. and A. Giazitzoglu (2019), "Youth, Enterprise and Precarity: Or, What is, and What is Wrong with, the 'Gig Economy'?", *Journal of Sociology*, **55** (4), 724–740.

Malhotra, A. and M. Van Alstyne (2014), "The Dark Side of the Sharing Economy… and How to Lighten It", *Communications of the ACM*, **57** (11), 24–27.

McGoogan, C. and J. Yeomans (2016), "Uber Loses Landmark Tribunal Decision over Drivers' Working Rights", *The Telegraph*, October. Available at: https://www.telegraph.co.uk/technology/2016/10/28/uber-awaits-major-tribunal-decision-over-drivers-working-rights/ (accessed on 26 September 2022).

Mexi, M. (2020a), "The Future of Work in the Post-Covid-19 Digital Era", Social Europe, April. Available at: https://socialeurope.eu/the-future-of-work-in-the-post-covid-19-digital-era (accessed on 26 September 2022).

Mexi, M. (2020b), "The Platform Economy – Time for Decent Digiwork", Social Europe, November. Available at: https://socialeurope.eu/the-platform-economy-time-for-decent-digiwork (accessed on 26 September 2022).

Oesch, D. (2010), "Trade Unions and Industrial Relations in Switzerland", MPRA Paper No. 22059. Available at: https://mpra.ub.uni-muenchen.de/22059/ (accessed on 26 September 2022).

Organisation for Economic Co-operation and Development (2019), *Negotiating Our Way Up: Collective Bargaining in a Changing World of Work*, Paris: OECD.

Organisation for Economic Co-operation and Development (2020), "The Future of Work", Expert Meeting on Collective Bargaining for Own-Account Workers Summary Report, Paris: OECD.

Piasna, A., W. Zwysen and J. Drahokoupil (2022), *The Platform Economy in Europe*, Working Paper 2022.05, ETUI, Brussels.

Prassl, J. (2018), "Collective Voice in the Platform Economy: Challenges, Opportunities, Solutions", Report to the ETUC.

Prassl, J. and M. Risak (2016), "Uber, Taskrabbit, & Co: Platforms as Employers? Rethinking the Legal Analysis of Crowdwork", *Comparative Labour Law and Policy Journal*, 37(3). Available at: https://papers.ssrn.com/sol3/papers.cfm?abstract_id=2733003 (accessed on 26 September 2022).

Risak, M. and J. Warter (2015), "Legal strategies towards fair conditions in the virtual sweatshop", Paper presented at the 4th Regulating for Decent Work Conference (ILO, Geneva, 8–10 July).

Roberts, Y. (2018), "The Tiny Union Beating the Gig Economy Giants", *The Guardian*, July. Available at: https://www.theguardian.com/politics/2018/jul/01/union-beating-gig-economy-giants-iwgb-zero-hours-workers (accessed on 26 September 2022).

Rueda, D. (2005), "Insider–Outsider Politics in Industrialized Democracies: The Challenge to Social Democratic Parties", *American Political Science Review*, **99** (1), 61–74.

Sabeel Rahman, K. (2016), "Reinventing the Social Contract" (draft). Available at: https://fdocuments.in/document/reinventing-the-social-contract-roosevelt-institu-reinventing-the-social-contract.html (accessed on 26 September 2022).

Schmidt, F. A. (2017), *Digital Labour Markets in the Platform Economy*, Bonn: Friedrich-Ebert Stiftung.

Smith, R. and S. Leberstein (2015), "Rights on Demand: Ensuring Workplace Standards and Worker Security in the On-Demand Economy", Washington: National Employment Law Project (NELP).

Soderqvist, F. (2017), "A Nordic Approach to Regulating Intermediary Online Labour Platforms", *TRANSFER – European Review of Labour and Research*, **23** (3), 349–352.

Standing, G. (2015), "Taskers: The Precariat in the On-Demand Economy (Part One)". Available at: https://workingclassstudies.wordpress.com/2015/02/16/taskers-the-precariat-in-the-on-demand-economy-part-one/ (accessed on 26 September 2022).

Stanford, J. (2017), "Historical and Theoretical Perspectives on the Resurgence of Gig Work", *Economic and Labour Relations Review*, 3 (28), 328–401.

Stewart, A. and J. Stanford (2017), "Regulating Work in the Gig Economy: What are the Options?", *The Economic and Labour Relations Review*, **28** (3), 420–437.

Tassinari, A. and V. Maccarrone (2020), "Riders on the Storm: Workplace Solidarity among Gig Economy Couriers in Italy and the UK", *Work, Employment and Society*, **34** (1), 35–54.

Taylor, M. (2017), "Good Work: The Taylor Review of Modern Working Practices", Department for Business, Energy and Industrial Strategy, London.

Available at: https://www.gov.uk/government/publications/good-work-the-taylor-review-of-modernworking-practices (accessed on 26 September 2022).

Trappmann, V., I. Bessa, D. Joyce, D. Neumann, M. Stuart and C. Umney (2020), *Global Labour Unrest on Platforms: The Case of Food Delivery Workers*, Trade Union in Transformation 4.0. Berlin: Friedrich-Ebert-Stiftung.

Valenduc, G. and P. Vendramin (2016), "Work in the Digital Economy: Sorting the Old from the New", ETUI Working paper No. 2016.03.

Vandaele, K. (2021), "Collective resistance and organizational creativity amongst Europe's platform workers: a new power in the labour movement", in J. Haidar and M. Keune (eds), *Work and Labour Relations in the Global Platform Economy*, Northampton, MA, USA and Cheltenham, UK: Edward Elgar Publishing, pp. 206–235.

Vaughan-Whitehead, D., Y. Ghellab and M. Munoz De Bustillo Lorente (eds) (2021), *The New World of Work, Challenges and Opportunities for Social Partners and Labour Institutions*, Northampton, MA, USA and Cheltenham, UK: Edward Elgar Publishing.

Wood, A. J., M. Graham and V. Lehdonvirta (2019a), "Good Gig, Bad Gig: Autonomy and Algorithmic Control in the Global Gig Economy", *Work, Employment and Society*, **33** (1), 56–75.

Wood, A. J., M. Graham, V. Lehdonvirta and I. Hjorth (2019b), "Networked but Commodified: The (Dis)Embeddedness of Digital Labour in the Gig Economy", *Sociology*. Available at: https://journals.sagepub.com/doi/full/10.1177/0038038519828906.

Zwick, A. (2018), "Welcome to the Gig Economy: Neoliberal Industrial Relations and the Case of Uber", *GeoJournal*, **83** (4), 679–691.

2. The gig economy and social partnership in Germany: towards a *German Model 4.0*?[1]

Johannes Kiess

1. INTRODUCTION

This chapter asks whether the gig economy brings about a "disruption" of social partnership in Germany and, consequently, leads to the restructuring of the *German model* of capitalism. It shows that, against such an assertion and beyond strong disagreement in policy reactions, most actors in Germany are confident in the relative stability of the institutional framework. Researching the implications of the gig economy in Germany, though, needs to be contextualized within a broader debate, especially given the general dominance of the industrial sector within the German corporatist arrangement (for example, Marsden 2015): digitalization and "industry 4.0" (Botthof and Hartmann 2015; Pfeiffer 2017) are indeed the most dynamic debate in German economic and labour policy (Pfeiffer and Huchler 2018). The Covid-19 pandemic is further accelerating this debate with regard to home office and the infrastructure needed for work in the digital world. The gig economy, I assume, serves as a symbol for the wider debate on disruptive technological change.

Moreover, this issue is embedded in the ongoing academic debate about processes of erosion and revitalization of the *German model* of capitalism (Anderson et al. 2015; Baccaro and Benassi 2017; Baccaro and Howell 2011; Holst and Dörre 2013; Kiess 2019b; Marsden 2015; Schulten 2019; Rothstein and Schulze-Cleven 2021; Unger 2015). This model rests not only on government institutions and ideology – that is, the tradition of the "social market economy" – but also on the "embedded" class conflict and, thus, the institutions of social partnership. Indeed, familiar patterns of social partnership and classic ideological divides structure the debate about how to organize digital change (state vs. social partners vs. the market: Kiess 2019a, 2019b; Kinderman 2014). While employers regard technological change as a mostly entrepreneurial challenge, trade unions are eager to moderate this change through social partnership and government intervention. However, in this new debate, trade unions are faced with no counterpart as platform businesses are reluctant to join existing organizations and arrangements, claiming they are not employers. Because the gig economy may increase the size of the low-wage sector, German social partnership might indeed be facing challenges of further dualization and precarization.

To investigate these challenges, I analyse how German actors define and operationalize the seminal notion "gig economy" in their strategic activities and how they describe the future of social partnership in the light of gig work. I rely on 22 semi-structured interviews conducted between the late summer of 2018 and spring 2019 with officials from trade unions, business associations, and government entities. In addition, we collected statements, policy papers, and reports published by these organizations between 2015 and 2019. With its focus on frames and ideas, the chapter takes an original perspective on industrial relations research. While the phenomenon "gig economy" is still very young and rapidly changing, the positions and frames adopted by central actors in Germany appear to be relatively stable so that this chapter's findings are instructive for the near future.

The following section provides a basic account of the *German model* characteristics and of the government and social partners' understandings of challenges that digitalization and particularly the gig economy pose, herewith presenting actors' different stories of the gig economy. Subsequently, I discuss three main issues that social partners see as affected by gig work: (a) the regulation of work and defining employer/ employee statuses; (b) access to and adequacy of social protection; and (c) prospects of social dialogue and collective action. Against this background, I then discuss whether and why actors seek cooperation and what we can learn from the empirical material about the future development of social partnership and the *German model* to a model 4.0. The chapter concludes with a summary and outlook on the asserted "disruption" of German social partnership induced by the gig economy.

2.　THE *GERMAN MODEL* AGAINST THE GIG WORK CHALLENGE

2.1　Incremental Change and the Stickiness of Established Institutions of Social Partnership

The institutionalized inclusion of social partners in politics is characteristic of the *German model* of capitalism. This important role of social partners together with jointly administered instruments like the short-time working scheme have gained increased relevance in recent economic crises (Herzog-Stein et al. 2013; Kiess 2019b; Schulten and Müller 2020). While collective bargaining coverage and union density have decreased over past decades (Vachon et al. 2016; Visser 2006), the coverage rate remains medium-high compared with other industrialized countries (56 per cent in 2017, according to the OECD). Another feature of the *German model* is the vocational training system that works as an incentive for long-term cooperation (Estevez-Abe et al. 2001), ensuring both employers and trade unions have a stable interest in cooperation. Even the subtle but continuous decentralization of collective bargaining taking place since the 1990s mostly occurs in an organized manner (Ibsen and Keune 2018). Last but not least, diversified quality production (Sorge and Streeck 2018) and the importance of manufacturing explain the continued importance of cooperation between social partners as well as the prominence of "industry 4.0" in the wider debate on digitalization.

The Agenda 2010 labour market reforms started in 2002 by the Social Democrat and Green coalition have led to the partial liberalization of social policy, increased pressure on the unemployed, and considerable political contestation (Bruff 2010; Hassel and Schiller 2010; Hegelich et al. 2011; Lahusen and Baumgarten 2010). Trade unions face pressure because of power and institutional imbalances (Rothstein and Schulze-Cleven 2021). Some authors even announced the "end of the conservative welfare state model" (Seeleib-Kaiser 2016) and proclaimed a decline of the coordinated model (Streeck 2010). Still, the change of German welfare and labour market institutions appears to be rather incremental (ibid.; Streeck and Thelen 2005) and more inclined to follow along existing path dependencies (Kemmerling and Bruttel 2006; Kiess et al. 2017; Seeleib-Kaiser 2016) than facing "disruptions" and large-scale dismantling of welfare institutions.

While there remains a strong social partnership in Germany, the literature always emphasized the confrontational nature of the relationship (Crouch 1982; Korpi 1974, 1983; Schmalz and Dörre 2014). German social partnership, therefore, has often been called conflictual partnership to underline the character of compromise between capital and labour. Hence, in this chapter, the debate on the gig economy and digitalization is read against the background of confrontational interests within and beyond the existing framework of the *German model*. Indeed, the traditional cleavage between trade unions and employers (and conservative/liberal and social-democratic/left parties) is clearly visible at least for the central and most powerful organizations: trade unions criticize newly emerging forms of employment as "exploitation 4.0", while employers warn against quick shots in regulation, which they perceive as emotion-driven. At the same time, perceptions and positions concerning digitalization are more complex. New conflicts also arise between different business models (Uber vs. taxis, hotels vs. Airbnb). Among labour representatives and different groups of workers, varying positions emerge. Besides, the debate is fragmented since actors struggle to position themselves in light of (perceived) uncertainties. For example, the interests of self-employed and freelancers are less clear than those of traditional workers regularly employed in the manufacturing industry. Last but not least, as Kirchner and Beyer (2016) argue, with the emergence of platforms we observe competition between the competitiveness logics of the "old" and "new" economy. This and the wide variety of platform business models may make coordination increasingly difficult.

The evaluation of all these developments differs widely, as do expectations of how they affect work, and, consequently, social partnership and the institutions of the *German model* of capitalism. While the pressure induced by digitalization already affects the discourse of social partners, actual change occurs incrementally and the institutions of the *German model* remain sticky (Streeck and Thelen 2005; Boettke et al. 2008). We contend that actors operate and negotiate within existing institutional boundaries and contribute to this stickiness by supporting institutions as long as they see them fit, and also renegotiate institutional settings in their favour if they deem it possible or necessary (Crouch 1982; Fehmel 2010, 2014; Kiess 2019a, 2019b). Thus, the "disruption" of traditional social partnership induced by the gig work phenomenon lies predominantly in the (differences of) "imagined futures" (Beckert 2016) of social partnership rather than in actual practices.

2.2   Digitalization as a Challenge? A Classic Conflict between Labour and Capital

While all interviewed public officials voiced general awareness concerning already existing precarious employment and potential risks of gig work, the government is not planning any immediate or concrete measures. This might be due to lack of ambition as labour representatives allege (see below) or to actual uncertainties as government officials claim, or even to the political logic of the grand coalition, which seems to limit the room of manoeuvre for more ambitious policies.[2] Government officials agree that there is a need to gather more information through commissioned research to inform legislative action. Officials regard dramatic predictions like the PWC study (PWC 2015), claiming an already huge impact of the sharing economy, as questionable. Also, an official of the Federal Ministry of Labour and Social Affairs (BMAS) described the field as ill-defined. However, the Ministry notices a general tendency that professions are decomposed into smaller tasks and the emergence of new, precarious jobs (for example, delivery services). The phenomena connected to gig work are seen as complex and diverse; regulation of only (some) platforms is difficult to implement and questionable from a regulatory perspective.

Yet, there is also contestation within the government. An official from the Federal Ministry for Transport and Infrastructure (BMVI, led by the conservative CSU), which is responsible for the digitalization strategy, pointed at the opportunities of digitalization if the regulatory framework is done right and stimulates innovation. As studies would show, the chances outweigh the risks for the labour market, especially if jobs are optimized and become more attractive. The view of the BMAS, led by the social-democratic SPD, deviates from this position, as the public siding of labour minister Hubertus Heil with food delivery riders in early 2018 illustrates. However, regulatory action would concern the "whole system" of German social and labour policy (see below), a scale of action that implies strong restrictions on what is politically feasible. This may also be the reason why the BMAS considers most important to understand what is happening and substantiate claims of precarization. This has led to considerable research commissioning. These varying government positions, oscillating between immediate regulatory action and letting the market decide, can roughly be identified with the classic left–right continuum: the Christian Democrats (in government) and especially the liberal FDP (in opposition) argue in favour of market liberalism, the Greens (in opposition) are optimistic about digitalization but have comparatively little interest in labour policy, the SPD (in government) favours careful and step-wise regulation, and the left (in opposition) strongly argues for structural reform and improving social protection. The right-wing extremist AfD does not have a clear position.

Towards the left side of this continuum, German trade unions take a decidedly proactive and prescriptive position concerning the gig economy and digitalization more generally. Trade unionists make a key differentiation between "low road" business strategies focusing on cost minimalization and a "high road" including service orientation, customer retention, high labour standards, and so on. The core claim is, "Digitalization must serve the people; platforms have to serve the people." Trade unions also explicitly challenge the "Silicon Valley way of digitalization", while they do recognize the potentials of mobile work in generating an income regardless of circumstances such as family obligations. For traditional IG Metall members in the automobile and similar sectors, the platform economy is not as threatening as competition from Eastern Europe. These perceptions limit targeting the gig economy, and the IG Metall concentrates on skills training and automatization – that is, the debate on industry 4.0 and labour 4.0. For the United Services Trade Union ver.di, though, platform-based business models make a difference because they reduce costs by shifting responsibility to employees of the service sector. Trade unions naturally take issue with the resulting precarization, summarized in a simple claim: "We don't need jobs that don't provide" – that is, jobs with too little income and protection. Since precarious employment is not new, especially in the service sector, ver. di tries to use the "hype" on gigging to improve regulations especially for the so-called solo-self-employed.

Fitting with the classic capital–labour cleavage, German business associations disagree fundamentally with trade unions and adopt a market-liberal perspective. While they claim expertise on evolving business models and digitalization, they are considerably less enthusiastic about engaging in consultation processes perceived as being too much about regulation and interference with market processes. While the media frequently report about delivery riders, clickworkers, and Uber, the phenomenon is, in the view of the Confederation of German Employers' Associations (BDA), still very limited. The shortage of skilled workers would additionally limit the future effects of technical possibilities to outsource since companies attempt to bind talent. Moreover, freelancers in IT and programming are portrayed as often earning well and themselves not interested in regular employment. Last but not least, across-the-board regulation would not address "bad platforms" effectively because business models differ extremely. At the same time, the vast and positive potential of platform work would be limited. Gesamtmetall (the German metal industry business association) further criticizes the BMAS' positions as defensive, presumptuous, and "not neutral" because it would not value the chances (and necessities) of a quasi-natural modernization process – that is, digitalization. Another central player, Bitkom, claiming to represent the digital economy, holds that platform workers themselves see many advantages in new forms of flexible work. Therefore, the debate around labour 4.0 should not be driven by the intention to define a new "standard employment relationship" (*Normalarbeitsverhältnis*) (Bitkom 2016b, p. 2). Rather, the government should ensure a "level playing field", only regulating where absolutely necessary (Bitkom 2016a, p. 3). All these arguments align with a general market-liberal perspective that regards government intervention, a necessity for trade unions, as generally problematic and digitalization as a quasi-natural process (for a critical perspective, see Rothstein 2021).

More direct engagement with the platform economy is observable in the hospitality sector where platforms significantly changed market rules. Accordingly, controversies occur not so much between social partners, but around the organization of markets. DEHOGA, the German Hotels and Hospitality Association, insists on "levelling the playing field" (since booking.com, expedia.com, and other platforms affect competition between hotels), not on the relation between hotels and their employees. As a result, business associations, not trade unions, tend to dominate these regulatory debates in attempts to defend their members' markets. At its annual convention in 2014, DEHOGA complained vehemently about increasing regulation, most importantly the introduction of a national minimum wage, while at the same time, the sharing economy would remain unregulated with negative effects for businesses and society (DEHOGA 2014). Following other business associations, with the notion "levelling the playing field", DEHOGA adopts a low-regulation, market-liberal ideology: "In this country, we have too much, rather than too little regulation."

German platform businesses have a slightly different story to tell: while their interests with regard to low and market-liberal regulation align with traditional business associations, they keep their distance from them and are not involved in the internal search for compromises of associations. This has several reasons: German platform businesses seem to concentrate on niche markets (brokering clickwork or non-stationary gig work) and, like global players, such as Amazon, eBay, and Airbnb, understand themselves as distinct from classic industries. Opposing perceptions of gig work replacing regular jobs, platforms often emphasize their workforce would consist mainly of students, parents responsible for taking care of their children, and other people not able to have a regular job and in need of flexibility. While the diversity of business models makes any regulation difficult, they argue new regulation for all platforms is not necessary since existing laws already apply to traditional and platform businesses alike. Underlying these claims from the platforms, there is the perception that their business models are on top of the regular economy, demanded by the market, and within the boundaries of existing frameworks. And since gig workers are not employed by them, there is no need for social partnership even if they wouldn't oppose it in principle. Hence, platforms engage in political consultation merely as observers.

In sum, while the policy debate on gig work is inconclusive on all aspects, established stakeholders invest considerable resources by commissioning research, drafting policy briefs, or setting up specialized units to cover the topic. However, they concentrate on the topics industry 4.0, labour 4.0, and digitalization rather than gig work in particular. Here, the government, so far, stays with observing the development, while the trade unions pressure for state intervention, and business associations warn against any quick-shot regulatory action. Most actors interviewed represent traditional sectors where gig work is interpreted rather as an external pressure than an ongoing trend, while platform businesses abstain from setting up social partnership institutions claiming they are not employers. The lobby impact of Uber, Facebook, and others on policymaking, was not covered by this study but, interestingly, it was not mentioned by any interview partner. While all actors agree that the topic is diverse and needs careful debate, this debate has not led to large-scale decisions.

3. HOW THE GIG ECONOMY CHALLENGES THE *GERMAN MODEL*

Social partners disagree on the consequences of digitalization and the need for action in particular. However, they concur that change is ongoing and inevitable. This section turns to three issues discussed in the interviews and fundamental for the *German model* of capitalism: (a) the regulatory framework of work and in particular how the emerging gig economy puts into question the status of workers; (b) the access to and adequacy of social protection, which touches fundamental questions of the German welfare state; and (c) the prospects of social dialogue and collective action.

3.1   The *Gretchenfrage* of Labour Rights: The Status of Gig Workers

The status of workers is a core question to any regulation of labour markets. Answering to disruptions caused by or expected from emerging platform businesses, state and business representatives – not trade unions – point at existing regulation that would already provide protection. This is not limited to labour law: one exemplary case is the taxi industry where the existing legislation effectively results in a ban for Uber. Hence, taxi drivers are protected, though not through new regulation of platform-brokered work. However, all interview partners acknowledge that the mere idea of gig work already challenges existing concepts of employer, employee, freelancer, firm, and so on. While these concepts and associated responsibilities are defined in German labour law, the proper classification of gig workers is debated (Childers 2017; Todolí-Signes 2017).

The term gig worker, most interview partners agreed, includes very different individuals in terms of education, income, work conditions, job autonomy, social security, and so on. Moreover, every platform, they assert, is a different case and, therefore, policymakers regard one-size-fits-all regulation as insufficient. Here, on the one hand, trade unions are concerned because of the experiences made with deregulation in the early 2000s that led to precarization especially among the solo self-employed (Bäcker and Schmitz 2016). Many existing worker protection regulations, anti-discrimination laws, parental leave, and other social policies are less favourable or not applicable to the self-employed. Growing numbers of workers with unclear statuses would further increase precarity. On the other hand, as all actors acknowledge, the definitions of employee and self-employed are related to social security dues and, consequently, eligibility for example to pensions (see below). As all actors have an interest in a reliable and accountable welfare state, the status question leads to pressure on policymakers to modernize labour law. Last but not least, there is a cultural element to questions of workers' statuses that for young people may have different meanings than for traditional blue- or white-collar workers.

The BMAS, although not the government coalition as a whole, sees the necessity to address the problems attached to the often-unclear status and weak protection of gig workers, arguing that the gig economy will grow. Many gig workers are indeed not covered by employee protection, as a researcher at the Federal Institute for Occupational Safety and Health assessed. Already, freelancers and the self-employed are mostly excluded from normally applicable worker protection and respective controls. Hence, gig work could lead to even more people working without protection, starting with the existing working time law, the risk assessment by companies, and including physical and psychological risk avoidance. For example, an employed craftsman has regular hours, regular income, social security, and safety equipment, all because the employer, by law, has a responsibility for the worker (*Fürsorgepflicht*). If the craftsman is self-employed and taking orders brokered via a platform, he has to take care of all that himself. Likewise, while at construction sites rules and controls exist, in essence, these are not applicable for self-employed workers. This leads to problematic risk-taking, be it because gig workers want a certain job, want to make (more) money, or simply because they have to pay their bills. Still, defining rules for different forms of gig work is difficult. In sum, while there is expertise within government bodies and a growing body of research, it seems that there is no clear policy idea. This is partly due to the fact that the issue is not relevant enough to top the political agenda of the coalition government.

For trade unions, though, regulating the platform economy is about crucial economic questions, about who sets what terms, and this concerns, above all, the status question. It may not be easy to force companies to accept their *Fürsorgepflicht* because of the global scale of the sector, but in the view of labour representatives, politicians often use this as an excuse. From a labour perspective, they argue, the case is obvious anyway: platforms are employers and only use their so-called new business models – contracting or brokering labour – to circumvent regulations. A local DGB secretary pointed at the example of Deliveroo[3] riders who elected a works council: the riders were discharged based on their temporary contracts and offered the opportunity to become freelancers, this time without the right to a works council. This type of employment is not legal in the view of trade unions, since a person riding six to eight hours per day for one platform will receive all orders channelled through that one platform. Thus, he or she is unable to accept or disregard orders or take on other work as a freelancer would, making this type of arrangement yet another case of ostensible self-employment.

Moreover, this concerns not only labour rights and the right to a works council but also, fundamentally, the national minimum wage, which does not apply to freelancers. Here, another problem is that existing competition law prevents self-employed workers from agreeing on prices. In other words, the self-employed cannot enter into collective bargaining as this would be considered a cartel. In the trade unions' view, however, the self-employed are a long way away from being at eye level with companies or ordering parties, which (potentially) results in underscoring the national minimum wage.

While the defining question for the worker status is whether s/he is dependent on one or a very small number of employers, moreover the trade unions struggle with the variety of jobs in the gig economy and the fact that the interests of and the circumstances for workers have changed. Unions try to keep up with such changes: since 2014, ver.di has organized yearly congresses on digitalization and it operates websites for crowdworkers. The service sector union also has a separate unit organizing and stewarding about 30,000 freelancers. Here, the status question is discussed openly and with a case-by-case approach. In ver.di's pragmatic approach, it is not important if the work is dependent or not as long as the platform pays social insurance dues and takes on its responsibility in monetary terms. There is, however, no mandatory obligation to date for the ordering party to pay social insurances (see below).

Following their general position on digitalization, employer representatives see no need for new regulation regarding the status of gig workers. Platform companies in particular insist that their relationship with gig workers is not one of employer and employee. The argument varies somewhat from case to case but generally points at the business model and the kind of service they provide, which, in their opinion, rests on brokering the workforce or service but not providing the service itself. Beyond that, existing regulations would already address the main problems. The interview partner from Bitkom acknowledged that food delivery riders were in ostensible self-employment, but this was in fact already illegal under current law. New regulation should only be considered after close examination. For now, the representative argued, platform workers themselves would be the last ones to think about further restrictions. Traditional employer associations like BDA, Gesamtmetall, and the associations representing the transport and hospitality sectors too, argue vehemently against touching existing definitions of employer, employee, and firm. Their idea of "levelling the playing field" then essentially means less regulation for traditional industries and employment as well, and gig work is a symbolic resource to achieve this goal.

The employers' anti-regulation position thus mirrors the trade unions' position: the latter develops the argument about precarious gig work around ostensible self-employment, the deregulation of the labour market in the early 2000s, and labour standards more generally. Against this, BDA, Gesamtmetall, and Bitkom argue that regulation is already excessive and that the gig economy does not change anything in principle. An argument often brought forward relates to the interests of gig workers themselves: traditional employment may not fit everyone and more flexibility is attractive to some segments of the labour force. Based on the conflicts between food delivery riders and their platforms (Deliveroo, Lieferando,[4] most recently Durstexpress), however, it is safe to assume that workers, too, differentiate between job flexibility and platforms' attempts to circumvent regulations.

3.2   Social Protection and the Social Security System

The second major issue raised by interview partners was mandatory social insurance contributions by and, consequently, the eligibility of gig workers and the self-employed for social security benefits. The literature characterizes Germany's social welfare system as conservative (Arts and Gelissen 2002; Esping-Andersen 1990; Kiess et al. 2017): contribution-based schemes favour those in standard employment, provide insiders with relatively high benefits, and partially exclude labour market outsiders. These welfare arrangements are an important pillar of the coordinated *German model* of capitalism producing legitimacy, stability, and the bases of continued coordination between social partners (Coates 2000; Hall and Soskice 2001; Amable 2003). Going beyond gig work, the social security debate includes discussions on reforming the present system to a more universal welfare state. While most interview partners agreed that the system should be adjusted to the modern world of work, the positions on how to reform the system remain incompatible. Employers stick to market-radical positions and are in support of lower labour costs and taxes in the face of global competition. Moreover, they favour private insurance schemes over a universal system. In contrast, labour representatives and left-wing politicians argue for universal coverage and making platforms responsible for social security payments. Again, as will become apparent in the following paragraphs, gig work rather impacts the *German model* as a symbolic resource – that is, the issue "gig economy" is used as an argument for changing existing arrangements and balances of power.

The coalition agreement between CDU/CSU and SPD, signed in March 2018, proposes to address the challenges associated with digitalization and globalization by developing new reporting standards, strengthening welfare state research, and further developing and supporting the New Quality of Work Initiative. The welfare state is to be modernized and adjusted for example, regarding the German Social Accident Insurance and the law on occupational illnesses. However, there is no agreement within the government to make the system (more) universal. Meanwhile, following an assessment by the IAB, a government-sponsored research entity, the digital transformation potentially harms the social welfare system as a whole, not just the coverage of individual workers: if a worker reduces her regular work and takes on additional jobs via platforms, she does not pay insurance dues and taxes on the latter. To compensate this, the state would need to ensure that platforms contribute to the health, care, and pension schemes directly. Accordingly, the left-wing opposition vividly criticizes the risks of underfunded pensions, limited health insurance, and lack of accident insurance. In their view, only supplementary regulation could include such irregular employees in the welfare system. However, since there is little political agreement, the phase of observation continues through 2021. At least for the pension system, a step-by-step adaption is observable (for example, the new *Grundrente*[5]) and there is an ongoing, controversial policy debate.

As a local DGB representative argued, these debates, especially about the pension system, will become more and more important because low-wage earners and gig workers never paid into the system but at the end of the day will receive at least social aid. Hence, the gig economy attacks essential pillars of the welfare state. The problem is multidimensional and touches not only the individual worker and his/ her rights, but the whole welfare state mechanisms: the state loses taxes because low-wage earners do not pay income taxes, gig workers (or their platforms, respectively) are not obliged to contribute to the schemes, regular businesses and workers are disadvantaged, and, in the end, tax money will be needed to support the poor. In the view of a freelancer organizer at ver.di, this situation is "prehistoric" since Germany is the last country in which the self-employed do not pay into the pension system. Since in Germany employers and employees pay the insurance dues in parity, wages themselves become an issue: many self-employed workers do not earn enough to be able to afford to pay both (the employer and employee) shares. Moreover, the market price difference for a given service provided by an employed person and a self-employed person (who gets tasks via a platform), the ver.di organizer calculated, is 20–21 per cent resulting in unfair competition. The trade unions thus suggest that platforms should automatically pay mandatory contributions like regular employers, regardless of the status of gig workers.

While the German employers' association (BDA) sees no need for additional protection of regular workers, it welcomes lower fees for self-employed (such as the already lower health insurance dues) and the mandatory inclusion of self-employed workers because people with very low incomes otherwise risk depending on social aid in old age. This is surprising given the general stance against regulation, but it can be explained by the interest of employers of not switching towards a universal, tax-financed welfare scheme. This would be the result, though, if the state had to step up social aid to prevent old-age poverty. Despite this potential agreement on the need to act to protect the existing social security system, at least Gesamtmetall criticizes the fact that left-wing parties would engage in clientele politics. In the view of the employers from the metal and electronics industries, any additional regulation is unnecessary given the small scope of the problem and because people are falsely deemed in need of help. These attacks are directed particularly at the SPD and the BMAS, which Gesamtmetall believes to be on a turn to the political left, and, thus, feed into the general ideological cleavage of the German political system.

In sum, while reforms are being considered and some measures have already been implemented (for example, lower health insurance dues for freelancers and the *Grundrente*), the government is reluctant to go beyond these small steps. Again, this needs to be seen in the light of the political climate in which the grand coalition makes major reforms not agreed upon in the coalition agreement almost impossible, but it also illustrates the government approach towards the gig economy: not too much regulation too soon. In theoretical terms, this confirms expectations that any changes in the established welfare system will occur, if at all, in the form of incremental and small steps. Furthermore, such change will occur most likely where social partners can align their interests. This is not the case regarding a full shift towards a universal welfare system, but it is more likely with regard to minor issues such as the mandatory inclusion of self-employed workers into the existing pension scheme.

3.3   The Future of Social Partnership

Last but not least, the emerging gig economy may challenge the established social partnership in its organizational foundations. Aspects of a changing landscape of social partnership discussed in the literature include the declining membership in associations and trade unions and coverage of collective bargaining (*Tarif-* and *Verbandsflucht*). Indeed, most platform businesses are not members of classic associations to begin with. Global players such as Amazon have no incentive to join business associations as they are powerful enough on their own. Some of the (small) German crowdwork platforms organize in their separate association, the Deutscher Crowdsourcing Verband, to articulate their interests but not to form social partnerships. As already mentioned, platforms do not conceive of themselves as employers, which already limits the opportunities for social partnership in the gig economy. At the same time, some German platforms think of themselves as "social partners" in that they are open for consultations, dialogue, and sometimes even collective agreements. Still, for most platform and many start-up businesses, social partnership culture is something they have to (be forced to) learn first, as a labour representative noted. This is to some extent also true for workers in the tech industry who have, in the experience of trade unions, different ideas on what to prioritize compared with the classic trade union agenda.

The points raised pose challenges, first and foremost, for trade unions. Their responses to digitalization include a considerable amount of research and pilot projects, internal debates about strategies but also the nature of platforms. Given the frequent "self-employed" status of workers in the hospitality sector, most importantly in the food delivery sector, but also in other industries, an important part of trade unions' work is legal advice and taking to labour courts. In this regard, the traditional fight against union busting and other employers' strategies is taken up with platforms just as with other companies. While ver.di has been organizing freelancers in the service sector for a long time, now also the IG Metall (traditionally mostly manufacturing industries) and the NGG (food industry and hospitality sector) accept freelancers as members. This diverges from the classical union idea of organizing dependent workers (while, of course, the "independence" of platform workers is disputed, see above). Moreover, since platform businesses are not organized, there is often no collective partner to negotiate with, effectively limiting negotiations (if they happen at all) to the company level. Finally, the platforms' reluctance to consider themselves employers makes the establishment of regular collective bargaining relationships unlikely.

Among the challenges for trade unions, according to interview partners, there is the need to modernize without losing their identity as collective workers' organizations. Classic organizing takes place on the shop floor where a political secretary connects with workers willing to organize. There is, from the perspective of local union representatives, no alternative to this approach although social media and particularly chat groups are helpful tools: online activities, too, need to bring people together in real life to develop power together. Worker mobilization is becoming increasingly difficult because people working at home or with short-time gigs have fewer opportunities to meet and build solidarity ties with each other. Since the core trade union work needs to be face to face, for example, by identifying (informal) meeting points of workers, trade unionists also believe in their classic power sources, most importantly legally sanctioned works councils to establish on-site space for further organizing. There have been first attempts to form works councils (Deliveroo, Foodora), but it is too early to tell whether this is a valuable strategy, particularly given the freelance status of many gig workers.

Especially interview partners on the local level insist that trade unions are more than just insurance companies against social risks faced by individual workers. As in a traditional factory, gig workers need to be willing to organize, not least due to unions' democratic self-understanding. A general problem is seen in that young people in particular are attracted by jobs in the gig economy promising certain freedom, while thinking less about social security. Interview partners criticized such Zeitgeist but also emphasized that young people would often quickly discover how problematic their jobs are. Still, the depoliticization and fragmentation of the public sphere hinder engagement – that is, trade unions feel generally disadvantaged in political debates – and the omnipresent economic pressure keeps people from joining volunteer ranks of unions. Against this, the idea of the "cloud and crowd project" was to establish a counterweight providing an online space where crowd workers could rate the platforms.[6] With such projects, IG Metall and ver.di not only aimed at supporting gig- and clickworkers but also assigned responsibility to the employers. Joint activities with the *Crowdsourcing Verband* resulted in an ombudsman mediating between platforms and crowdworkers. Such emerging "social partnership" based on the goodwill of some of the Germany-based platforms, however, cannot match the established powerhouses in traditional sectors. Regulation of the gig economy could start with a code of conduct such as the one agreed between ver.di, IG Metall, and the *Crowdsourcing Verband*, setting at least transparency and communication rules. Such voluntary rules and further regulations, however, would eventually need to be enforced by the state to enforce minimum standards for the self-employed to counter non-participation of many platforms.

In sum, trade unions face many challenges associated with the gig economy: postmaterialist priorities of workers, spatial dissolution of work, the anonymity of social media, reluctance to allocate resources from traditional to new sectors with uncertain gains, legal restrictions for freelancers to form coalitions, and so on. Some of these challenges call for new mobilizing strategies; others, however, lead trade unionists to point at contextual circumstances and the role of the state. Employers are reluctant to engage in social partnership in the gig economy, be it because associations do not represent platforms or because platforms do not consider themselves employers. Only successful union organizing and consequent pressure will change the current situation with regard to social partnership. This is all the more true since trade unions themselves do not want state intervention in the principle of *Tarifautonomie*. Intervention to secure basic norms of worker protection also for self-employed and gig workers, however, is on top of the agenda.

4.   THE HIGH ROAD VERSUS THE LOW ROAD: CONTESTING THE *GERMAN MODEL*

Attention towards the gig economy in Germany started to develop only from 2014 onwards and thus was rather late, compared with the debate in the United States (Greef and Schroeder 2017). Explicit references in the political arena are still scarce and scattered across a range of policy fields, and legislative proposals. Moreover, the concept itself (like other buzzwords often used synonymously) covers very different labour market phenomena; its boundaries are nebulous and prospects uncertain. Many issues could be studied, focusing on precarious employment (Crouch 2019) instead of a specific form of labour contract (the "gig"), which remains unclear in terms of scope. Indeed, one is left to wonder whether classic freelancing is not already "gigging". Some actors emphasize "platform-brokered work" as a more accurate definition, but still describe the phenomenon and its size as opaque.[7] While all actors believe that digitalization is changing the labour market somewhat, the debate on the gig economy remains difficult to separate from the sharing economy including more than paid labour, cloud work (non-stationary but principally including regular employment), and already existing freelancing or solo self-employment.

It is important to consider the impact of the gig economy on social partnership in light of the institutional characteristics of the *German model* of capitalism. Part of this model of capitalism is a relatively high level of labour protection for workers in regular employment, widely accepted coordination between state and social partners in particular in times of crises or regarding fundamental questions (again, for well-organized core industries), the system of vocational training, and the "conservative" welfare regime. This setting is also reflected in the self-esteem of the trade unions who consider themselves important, up-to-date partners and problem-solvers. On the local level, though, and particularly in East Germany, interviews reflected more scepticism compared with the national-level trade union representatives.[8]

I have shown that social partners and the government alike indeed see the challenges of the gig economy to the *German model* as being limited by existing institutions and regulations. However, expectations for the future align with the traditional cleavage between labour and capital: trade unions fear further precarization following their relative weakening in the last decades as well as the liberalization and precarization induced through the Agenda 2010 reforms. These reforms have already produced a considerable low-wage sector that, the trade unions fear, could grow with an increasing number of insecure jobs in the gig economy. Hence, the already existing dualization of the labour market (Palier 2012; Palier and Thelen 2010) and the future of the German welfare state are on top of the agenda of social-democratic policymakers and trade unions. Both tend to address these issues as a choice between a coordinated and a liberal model of capitalism. Employers disagree with such accounts and point at the potential of the gig economy. Moreover, market-liberal economic ideas, including the enthusiasm for Silicon Valley-style platform-based business models (Rothstein 2021), continue to impact the ongoing transformation of the *German model*. Partly, as the employer side argues, the gig economy debate feeds into existing debates and arguments that the German economy needs to remain competitive and open to innovation, and to embrace technological change. Partly, as trade unions emphasize, the gig economy exacerbates the situation of workers in the service sectors where low-wage jobs and questionable working conditions are already widespread. The gig economy then is also an "imagined future" (Beckert 2016), implying imperatives for action to avoid or achieve a certain outcome and, thus, plays an important role as symbolic resource in the repertoires of stakeholders.

Beyond this ideational aspect, the interview partners were aware of the *potential* disruptions of the unregulated impact of the platform economy. The account of Kirchner and Beyer (2016), that the platform logic is a challenge to existing market organization across varieties of capitalism, reflects this much more alarming perspective. Platform businesses like eBay, Amazon, or Uber do not simply enter a market (in which other players then could react to the challenger by adjusted offers and pressure on the state to regulate); they fundamentally restructure and often strive to gain a monopoly position on the market. This dimension is mentioned in most interviews we conducted as well as in many public statements analysed. Although platforms have not reached the core of the German labour market, trade unions and (some) social-democrats refer to such fundamental challenges when they speak of a "high-road" strategy for developing a "social market economy 4.0". In the words of former Labour secretary Brigitte Zypries (SPD), the challenge is to develop a "social market economy 4.0" to ensure that platforms align with the social standards and values of society (Der Spiegel 2017, p. 31). The BMAS' Whitebook also speaks of a "third way" between the current *laissez-faire* in the USA and state-led modernization in China (ibid., p. 32). This path should include observing trends, acting carefully, "and, wherever possible, favour social compromise with social partners and other actors of the labour market" (BMAS 2017, p. 188, author's translation).

Employers' associations deem such claims unrealistic and/or consider that they put the competitiveness of the German economy in jeopardy. Moreover, they claim that the negative effects of the gig economy on workers are exaggerated, while the potentials of platforms would deserve more mention. As they continue to participate in the consultation processes characteristic of the *German model*, business and employers' associations argue for "levelling the playing field" in the direction of low regulation and criticize particularly the BMAS for its endeavours. Social partnership, as other studies have shown before, is under constant pressure induced by global competition and employers' belief in market-liberal ideas. The gig economy, it appears, is merely the most recent playing field of this ongoing conflict.

Thus, against the talk of "disruptions", be they "inevitable", "dangerous", or "desirable", this chapter emphasizes two key findings for the *German model* and its prospect in the digital world. First, the *German model* is slowly adapting to the challenges of digitalization (investment in digital infrastructure, increased automatization, and so on) and changing cultural patterns of work–life balance (more flexible work, home office, even the adoption of universal elements into the conservative welfare state regime). Secondly, however, many of these changes are not specifically related to digitalization and only some have an explicit link to the gig economy (for example, preferences for a healthy work–life balance). Moreover, the changes are incremental and path-dependent instead of translating the business models of some platforms into "disruptive" change. While this chapter did not set out to present a thorough social policy or institutional analysis, this is reflected clearly in the accounts given by our interview partners and the public statements, policy briefs, and working papers. For some actors, the incremental change of the *German model* may not go far enough – trade unions wish for more and tighter social protection; business associations claim such regulation would inhibit necessary change. While for trade unions it is about resisting pressure for deregulation, others fear that any further regulation would put Germany even more behind the (supposed) Chinese and American digital "success stories". Employers (and platforms), nevertheless, will find it difficult to unilaterally push for levelling the playing field on such a "low road" model of capitalism.

NOTES

1. Stephanie Schneider contributed to the field work and Elisabeth Kissler helped with the literature research. I also thank Christian Lahusen for his valuable support and important comments on earlier drafts as well as an anonymous reviewer for his/her comments.
2. One interview partner claimed that a coalition crisis in summer 2018 over the refugee policy paralysed internal decision-making processes.
3. The UK-based food delivery company Deliveroo ceased all operations in Germany as of 16 August 2019 and now focuses on more lucrative markets.
4. Lieferando is the German branch of the Dutch Just Eat Takeaway food delivery company, which gobbled up the German brands pizza.de, foodora.de, lieferservice.de, and lieferheld.de. The market of delivery platforms keeps on being reorganized, as the example of beverage deliveries by *Durstexpress* shows.
5. From 2021 onwards, pensioners receive under certain conditions a higher basic pension (*Grundrente*), a reform that targets in particular the effects of irregular employment and the punishing effect of taking time off for (child) care.
6. For more information see http://cloud-und-crowd.de (last accessed 4 January 2022).

7. Even the most visible case in Germany, food delivery, is less clear since Foodora switched from freelance riders taking on separate "gigs" to (temporary) employment of riders.
8. In regions and sectors with relatively low collective agreement coverage and low union density, activities focus less on influencing policymakers and more on daily struggles. However, regional and sectoral differences should always be kept in mind when assessing the *German model* of social partnership (for an overview see Schulten 2019).

REFERENCES

Amable, B. (2003), *The Diversity of Modern Capitalism*, Oxford/New York: Oxford University Press.

Anderson, K., M. Baethge, and D. Sadowski (2015), "Editorial", *Journal for Labour Market Research*, **48** (2), 75–79.

Arts, W. and J. Gelissen (2002), "Three worlds of welfare capitalism or more? A state-of-the-art report", *Journal of European Social Policy*, **12** (2), 137–158.

Baccaro, L. and C. Benassi (2017), "Throwing out the ballast: growth models and the liberalization of German industrial relations", *Socio-Economic Review*, **15** (1), 85–115.

Baccaro, L. and C. Howell (2011), "A common neoliberal trajectory: the transformation of industrial relations in advanced capitalism", *Politics & Society*, **39** (4), 521–563.

Bäcker, G. and J. Schmitz (2016), *Atypische Beschäftigung in Deutschland*, Expertise für die Kommission "Arbeit der Zukunft", Duisburg-Essen: Institut Arbeit und Qualifikation.

Beckert, J. (2016), *Imagined Futures: Fictional Expectations and Capitalist Dynamics*, Cambridge, MA: Harvard University Press.

Bitkom (2016a), *Stellungnahme. Kommentierung Grünbuch Digitale Plattformen*, https://www.bitkom.org/sites/default/files/file/import/20161014 -Stellungnahme-Gruenbuch-Bitkom.pdf (14 October) (accessed on 22 September 2022).

Bitkom (2016b), *Thesenpapier. Die Deutsche Arbeitswelt Zukunftsfähig Gestalten*, https://www.plattform-i40.de/IP/Redaktion/DE/Downloads/ Publikation/bitkom-thesenpapier-arbeit-40.pdf?__blob=publicationFile&v=4 (29 September) (accessed on 22 September 2022).

BMAS (2017), *Weissbuch Arbeiten 4.0*, Bundesministerium für Arbeit und Soziales, https://www.bmas.de/SharedDocs/Downloads/DE/Publikationen/ a883-weissbuch.pdf;jsessionid=1CAC864920018801DE39973A512283A7 .delivery2-master?__blob=publicationFile&v=1 (November) (accessed on 22 September 2022).

Boettke, P. J., C. J. Coyne, and P. T. Leeson (2008), "Institutional stickiness and the new development economics", *American Journal of Economics and Sociology*, **67** (2), 331–358.

Botthof, A. and E. A. Hartmann (eds) (2015), *Zukunft der Arbeit in Industrie 4.0*, Berlin: Springer.

Bruff, I. (2010), "Germany's Agenda 2010 reforms: passive revolution at the crossroads", *Capital & Class*, **34** (3), 409–428.

Childers, R. (2017), "Arbitration class waivers, independent contractor classification, and the blockade of workers' rights in the gig economy", *Alabama Law Review*, **69** (2), 533–560.

Coates, D. (2000), *Models of Capitalism: Growth and Stagnation in the Modern Era*, Cambridge/Malden, MA: Polity Press.

Crouch, C. (1982), *Trade Unions: The Logic of Collective Action*, London: Fontana Paperbacks.

Crouch, C. (2019), *Will the Gig Economy Prevail?*, Cambridge: Polity Press.

DEHOGA (2014), *DEHOGA–Präsident Appelliert an Politik: "Mehr Unternehmerlust – Weniger Bürokratiefrust"*, https://www.dehoga -bundesverband.de/fileadmin/Startseite/06_Presse/Pressemitteilungen/ 2014/PM_14_22_DEHOGA-Branchentag_Praesident_Fischer_ _Mehr_Unternehmerlust_-_weniger_Buerokratiefrust_.pdf (accessed on 22 September 2022).

Der Spiegel (2017), Der dritte Weg, *Der Spiegel*, 2017 (12), 30–36.

Esping-Andersen, G. (1990), *The Three Worlds of Welfare Capitalism*, Cambridge: Polity Press.

Estevez-Abe, M., T. Iversen, and D. Soskice (2001), "Social protection and the formation of skills: a reinterpretation of the welfare state", in P. A. Hall and D. Soskice (eds), *Varieties of Capitalism*, Oxford: Oxford University Press, pp. 145–183.

Fehmel, T. (2010), *Konflikte um den Konfliktrahmen: Die Steuerung der Tarifautonomie*, Wiesbaden: VS Verlag für Sozialwissenschaften.

Fehmel, T. (2014), "Konflikte erster und zweiter Ordnung in Europa", *Leviathan*, **42** (1), 115–136.

Greef, S. and W. Schroeder (2017), *Plattformökonomie und Crowdworking: Eine Analyse Der Strategien und Positionen Zentraler Akteure*, Forschungsbericht 500, Kassel: Universität Kassel.

Hall, P. A. and D. Soskice (2001), *Varieties of Capitalism: The Institutional Foundations of Comparative Advantage*, Oxford/New York: Oxford University Press.

Hassel, A. and C. Schiller (2010), *Der Fall Hartz IV: Wie es zur Agenda 2010 kam und wie es weitergeht*, Frankfurt am Main: Campus.

Hegelich, S., D. Knollmann, and J. Kuhlmann (2011), *Agenda 2010: Strategien, Entscheidungen, Konsequenzen*, Wiesbaden: VS Verlag für Sozialwissenschaften.

Herzog-Stein, A., G. A. Horn, and U. Stein (2013), "Macroeconomic implications of the German short-time work policy during the Great Recession", *Global Policy*, **4**, 30–40.

Holst, H. and K. Dörre (2013), "Revival of the "German Model"? Destandardization and the new labour market regime", in M. Koch and M. Fritz (eds), *Non-Standard Employment in Europe*, Basingstoke: Palgrave Macmillan, pp. 132–149.

Ibsen, C. L. and M. Keune (2018), *Organised Decentralisation of Collective Bargaining: Case Studies of Germany, Netherlands and Denmark*, OECD Social, Employment and Migration Working Papers 217, 4 September.

Kemmerling, A. and O. Bruttel (2006), "'New politics' in German labour market policy? The implications of the recent Hartz reforms for the German welfare state", *West European Politics*, **29** (1), 90–112.

Kiess, J. (2019a), "Contention in times of crisis: British and German social actors and the quest of framing capitalism", in J. Kiess and M. Seeliger (eds), *Trade Unions under the Pressure of European Integration: A Question of Optimism and Pessimism?*, London: Routledge, pp. 208–226.

Kiess, J. (2019b), *Die soziale Konstruktion der Krise: Wandel der deutschen Sozialpartnerschaft aus der Framing-Perspektive*, Weinheim: Beltz Juventa.

Kiess, J., L. Norman, L. Temple, and K. Uba (2017), "Path dependency and convergence of three worlds of welfare policy during the Great Recession: UK, Germany and Sweden", *Journal of International and Comparative Social Policy*, **33** (1), 1–17.

Kinderman, D. (2014), *Challenging Varieties of Capitalism's Account of Business Interests. The New Social Market Initiative and German Employers' Quest for Liberalization, 2000–2014*, MPIfG Discussion Paper 16, Köln: Max-Planck-Institut für Gesellschaftsforschung.

Kirchner, S. and J. Beyer (2016), "Die Plattformlogik als digitale Marktordnung", *Zeitschrift für Soziologie*, **45** (5), 324–339.

Korpi, W. (1974), "Conflict, power and relative deprivation", *The American Political Science Review*, **68** (4), 1569–1578.

Korpi, W. (1983), *The Democratic Class Struggle*, London: Routledge.

Lahusen, C. and B. Baumgarten (2010), *Das Ende des sozialen Friedens? Politik und Protest in Zeiten der Hartz-Reformen*, Frankfurt am Main: Campus.

Marsden, D. (2015), "The future of the German industrial relations model", *Journal for Labour Market Research*, **48** (2), 169–187.

Palier, B. (2012), "Turning vice into vice: how Bismarckian welfare states have gone from unsustainability to dualization", in G. Bonoli and D. Natali (eds), *The Politics of the New Welfare State*, Oxford: Oxford University Press, pp. 232–255.

Palier, B. and K. A. Thelen (2010), "Institutionalizing dualism: complementarities and change in France and Germany", *Politics & Society*, **38** (1), 119–148.

Pfeiffer, S. (2017), "Industrie 4.0 in the making: discourse patterns and the rise of digital despotism", in K. Briken, S. Chillas, and M. Krzywdzinski (eds), *The New Digital Workplace: How New Technologies Revolutionise Work*, Wiesbaden: Springer, pp. 21–41.

Pfeiffer, S. and N. Huchler (2018), "Industrie 4.0 konkret – vom Leitbild zur Praxis?", *WSI-Mitteilungen*, **71** (3), 167–173.

PWC (2015), *The Sharing Economy*, Consumer Intelligence Series, https://eco.nomia.pt/contents/documentacao/pwc-cis-sharing-economy-1-2187.pdf (accessed on 22 September 2022).

Rothstein, S. (2021), "Toward a discursive approach to growth models: social blocs in the politics of digital transformation", *Review of International Political Economy* (online first).

Rothstein, S. and T. Schulze-Cleven (2021), *Imbalance: Germany's Political Economy after the Social Democratic Century*, New York: Routledge.

Schmalz, S. and K. Dörre (2014), "Der machtressourcenansatz: ein instrument zur analyse gewerkschaftlichen handlungsvermögens", *Industrielle Beziehungen: Zeitschrift für Arbeit, Organisation und Management*, **21** (3), 217–237.

Schulten, T. (2019), "German collective bargaining: from erosion to revitalisation?", *WSI-Mitteilungen*, **Special Issue 2019**, 11–30.

Schulten, T. and T. Müller (2020), *Kurzarbeitergeld in Der Corona-Krise. Aktuelle Regelungen in Deutschland Und Europa*, Policy Brief WSI 38, Düsseldorf: WSI.

Seeleib-Kaiser, M. (2016), "The end of the conservative German welfare state model", *Social Policy & Administration*, **50** (2), 219–240.

Sorge, A. and W. Streeck (2018), "Diversified quality production revisited: its contribution to German socio-economic performance over time", *Socio-Economic Review*, **16** (3), 587–612.

Streeck, W. (2010), *Re-Forming Capitalism: Institutional Change in the German Political Economy*, Oxford; New York: Oxford University Press.

Streeck, W. and K. Thelen (2005), "Introduction: institutional change in advanced political economies", in W. Streeck and K. A. Thelen (eds), *Beyond Continuity: Institutional Change in Advanced Political Economies*, Oxford: Oxford University Press, pp. 3–39.

Todolí-Signes, A. (2017), "The end of the subordinate worker? The on-demand economy, the gig economy, and the need for protection for crowdworkers", *International Journal of Comparative Labour Law and Industrial Relations*, **33** (2), 241–268.

Unger, B. (2015), *The German Model: Seen by Its Neighbours*, Falkensee: SE Publishing.

Vachon, T. E., M. Wallace, and A. Hyde (2016), "Union decline in a neoliberal age: globalization, financialization, European integration, and union density in 18 affluent democracies", *Socius: Sociological Research for a Dynamic World*, **2**, https://journals.sagepub.com/doi/full/10.1177/2378023116656847 (accessed 4 September 2022).

Visser, J. (2006), "Union membership statistics in 24 countries: an analysis of 'adjusted' Union membership data in 24 countries", *Monthly Labor Review*, **129** (1), 38–49.

3. Social partnership and the rise of the gig economy in Greece: continuity or discontinuity?

Maria Mexi

INTRODUCTION

Within the overall context of the emergence of the gig economy and the challenges that come with it (see Introduction in this book), the present chapter seeks to add to the literature on the intersection of the gig economy and social partner responses at national level. It particularly examines the impact of the Greek model of social partnership on actors' responses to the rise of the gig economy, and the role of employers and trade unions in Greece. It thus assesses the role of policy legacies and the extent to which they affect social partner responses and changes, and in what ways. Applying a modified version of the classification used by Hall (1993), we provide evidence on two types of impact. First, *first-level (ideational/cognitive) change*, which leads to an alteration of public policy discourses. In this regard, we identify how social partners perceive risks and opportunities in relation to gig economy and work, and whether and how they modify their language in discussing and analysing issues. This also involves identifying impacts in terms of knowledge diffusion and the broadening of the social partners' policy agenda, strategies, and tactics of mobilisation and representation. Second, there is *second-level (institutional/policy) change*, which opens up space for reforms in social partners' preferences and organisational resources. At this level of analysis, we have sought to identify whether and how social partners become actively involved in policy reforms to address growing problems related to the gig economy and labour, triggering institutional change through a change in their sets of preferences. It should be stressed that first- and second-level changes do not work in silos, but they are interwoven, as ideas reinforce the drive towards the adoption of policies.

Our research is based on the triangulation of qualitative research methods, namely a review of relevant policy and regulatory documents, an archival search of trade unions' and employers organisations' announcements, press releases and other printed and online material, and 28 in-depth semi-structured interviews conducted between mid-2018 and mid-2019 in Greece with policymakers, members of trade unions and employers' organisations, owners of digital platforms in the transport sector, policy experts, academics, journalists, and workers engaged in activities provided by or through digital platforms such as Uber and Airbnb.

The chapter is divided into four main sections. The first section offers a brief overview of the rise of the gig economy in Greece. The second section then detects and empirically assesses *first-* and *second-level changes*, highlighting key ideational, organisational (in terms of mobilising and representing actors in the gig economy) and policy responses. The link between the model of Greek social partnership and the gig economy is explored in section 3 as well as the salience of mediating variables, such as past legacies affecting Greek actors' responses to the gig economy challenge, which are examined in more detail in the fourth section. The chapter concludes by summarising the findings of our study and making suggestions for further action.

1.   THE GIG ECONOMY IN GREECE: AN OVERVIEW

Since 2009, Greece has become the epicentre of a series of crises with considerable socio-economic and humanitarian repercussions: the economic and unemployment crisis, the Eurozone crisis, and, more recently, the migration/refugee crisis. In terms of economic impacts, over the early crisis years, Greece experienced a sharp decline in its GDP from 241,990.4€ million in 2008 (when the economic crisis first started) to 176,022.7€ million in 2015.[1] At the same time, the Greek unemployment rate rapidly increased from 7.8 per cent in 2008 to 24.9 per cent in 2015,[2] while youth unemployment reached almost 50 per cent. It should be mentioned that the unemployment figures obscure the strikingly high unemployment levels among people with disabilities, which was more than double the national jobless rate of 23 per cent (ANED 2015/2016). "The sudden growth in unemployment," Visvizi (2016, para. 8) argues, "followed by sudden loss in disposable income level, and accompanied by a disintegrating state administration means that no social provision exists for those in need; and the numbers are growing. The private sector, swamped by excessive taxation, operating in an inflexible labour market framework, under conditions of a liquidity squeeze, cannot absorb the unemployed. Therefore, as the crisis continues, amidst political instability at home and abroad, the resources at the disposal of families dwindle. In this view, the degree of social deprivation is bound to increase." Overall, as fiscal consolidation measures were the primary priority for Greek policymakers throughout the entire crisis period following policy directions of the relevant memoranda of understanding signed with the country's international creditors, limited attention was paid to calls for targeted social measures to cater to those vulnerable and most affected by the country's multiple crises (Mexi 2018).

These negative effects contributed considerably to the growth of services provided by platforms such as Airbnb and, more broadly, to the expansion of the gig economy or "sharing economy", as it provided opportunities for employment and new sources of income. As one of our interviewees (representative of an employers' organisation) stated, "in a country with one of the highest unemployment rates in the European Union, digital platforms especially in the hospitality and tourism industry are creating jobs and new revenue opportunities." Yet, even though use of digital platforms experienced a significant growth, there was limited attention in the public discourses of what exactly these processes entailed and what were the most relevant multifaceted drivers and effects. Contrary to what was happening in other European countries, the gig economy and work were poorly understood, particularly in the realm of labour issues. For instance, unlike the policy debate and focus in other European countries from 2016 onwards, fundamental questions such as whether a gig worker can rightfully be classified as an independent contractor remained largely untackled in Greece. As emphatically pointed out by an interviewed trade unionist referring to the phenomenon of the gig economy, "This new sector is still uncharted waters for us."

By contrast with the international discourses and key debates on gig worker status, as well as employment rights and social dialogue in the gig economy milieu (see Introduction in this book), the gig economy in Greece did not develop as a particular topic of high contention between Greek workers and employers; and unlike other European countries, up to date, it has not given rise to a new framework for facilitating discussion, collaboration or social dialogue between the parties involved. As we explain below, several factors may have contributed to this result. One may be the relatively small size of the gig economy workforce vis-à-vis the general workforce, though the growing gig economy activities especially in the hospitality and tourism industries have the potential to rapidly change the quality of jobs and to completely reshape the business activities pertaining to those particular industries. Moreover, looking deeper, we have sought to assess how the nature and evolution of Greek social partnership affected social partners' responses to the gig economy in the early phases of its development and their capacity to instigate substantial policy reforms. Our empirical findings show an overall strong link between the prevailing modalities of social partnership and social partners' responses. This link is mediated by certain historical legacies and inherent weaknesses (or idiosyncrasies) of the Greek model of social partnership and trade unionism that proved difficult to reverse. In particular, Greece is a representative case of a Southern European country characterised by crony capitalism and weak labour market institutions, strong clientelism and low levels of policy concertation, a guild-oriented social structure and a history of adversarial industrial relations, in addition to trade union fragmentation and low institutionalisation of bargaining procedures (Zambarloukou 2006). All these elements are detailed in the next section, with a specific focus on the hospitality and transport sectors.

2.   THE GREEK CASE IN A CLOSER LOOK: KEY FIRST- AND SECOND-LEVEL CHANGES IN THE FIELDS OF HOSPITALITY AND TRANSPORT

To begin with, it should be emphasised that, in terms of first-level (ideational and cognitive) changes, there is evidence of a lack of attention to the concepts of gig economy and work in Greek public discourses during the early phase of its rise and growth. This is to be contrasted with the international experience, which shows that the gig economy has been subject to significant amounts of political discussion and contention in Europe and elsewhere, especially at its early phases of rise and growth. Airbnb attracted most attention in Greece due to its prevalence and size, followed by Uber (which had to suspend most of its service) and the ride-hailing app Beat (the almost Greek equivalent platform to Uber) in the transport sector. But public attention to these individual platforms came far away from an in-depth realisation and understanding of what the gig or sharing economy actually is and what it entails in terms of economic and labour market repercussions. Below, the specific cases of Airbnb and Uber are presented, looking particularly at the particular policy and regulatory responses observed at the second-level analytical focus.

2.1   Policy and Legislative Changes: The Case of Airbnb

Faced with rising unemployment and shrinking incomes (plus overtaxation of private property as part of the set of measures imposed on Greece by foreign creditors), a considerable part of the Greek population (homeowners) turned to digital platforms, such as Airbnb, to make a living or for pocket money (Goranitis 2016). Due to its popularity, Airbnb soon became an emblematic case of the gig or sharing economy in Greece. During 2017, 1,370,000 tourists who visited Greece had booked a place on Airbnb. Based on Inside Airbnb data (see Poutetsi 2018), there were 5,127 listings in Athens at an average nightly rate of 55€. About 4,268 or 83.2 per cent of these involved entire houses or apartments, 15.8 per cent (808) private rooms and only 1 per cent (51) shared rooms. As the data show, "Airbnbing" became a widespread and lucrative economic activity for many Greeks. In particular, according to a study conducted by the Athens University of Economics, in Athens Airbnbing brought approximately €69 million to individuals throughout the city. About 73 per cent of Airbnb's hosts are not employed in standard jobs, while 28 per cent state that the income coming from Airbnbing is a good supplement. A case study on Airbnb in Athens, involving in-depth interviews with 15 hosts, shows that financial resources are the main motives for host engagement (Lemonis 2015); yet the study also finds that hosts resorted to Airbnbing not only as a financial opportunity but also as an opportunity to socialise with and befriend international travellers as well as explore Athens in a way they had not done before (Rinne 2015).

Besides being an important source of income for individual Greeks during hard economic times, Airbnb grew into an "ecosystem" or a marketplace of supporting companies. It has thus contributed to opening business opportunities for smaller platform or physical companies (e.g., easyBNB, FlipKey, HouseTrip, Home Rentality and Vrbo),[3] which act as intermediaries between guests and hosts of Airbnb or provide services that Airbnb does not. These include cleaning services, transportation services (e.g., Hopwavecustomer), checking, property management and even pricing advice for property owners (e.g., Discoveroom). These platforms created new jobs especially for people out of work.

Yet alongside new opportunities, Airbnb presented new threats. It was thus seen as a prime example of a market disruptor, in that it has considerably transformed the market for accommodation in several Greek cities (Guttentag 2013), rivalling traditional industries such as hotels and real-estate agencies. Moreover, as an increasing number of properties are now being used as Airbnb accommodation, there are not enough left for residents. This resulted in increasing the cost of living in Athens, as the examples of other cities across the globe (Barcelona, New York, etc.) also show (Badcock 2017; Gutiérrez et al. 2017; Santolli 2017).[4] Overall, while supporters of such disruptive innovations argue that Airbnbing can yield positive outcomes, such as the empowerment of ordinary people as well as efficiency and lower environmental costs, critics attack digital platforms like Airbnb for being more about economic self-interest than sharing; and for being exploitative and predatory (Schor 2011, 2014). Along these lines, there are those who argued that Airbnb in Greece, as in other countries, is increasingly creating a new informal economy of uninsured workers, as entire professions – cab drivers, cleaners, etc. – are passing into a new grey market, unregulated, tax free and uninsured.

As our research has found, Airbnb's growing commercialisation and popularity triggered both first- and second-level changes during the early phases of the rise of the gig economy in Greece. These include: hostile rhetoric on the part of the traditional social partner organisations in the hotel industry; new laws to regulate the sector in light of growing tax avoidance by Airbnb hosts; and new opportunities for mobilisation and interest representation among Airbnb hosts as well as professional groups of people with similar interests.

More particularly, media reports, as well as lively debates at hotel, travel and tourism conferences, kept the spotlight on Airbnb's growth and success in Athens and several Greek cities in parallel with accusations of tax evasion being levelled at both Airbnb (the platform) and property owners. Concerning first-level impacts, media and policy discourses in Greece particularly focused on the unequal tax treatment between hotels and residential properties rented for the short term and how this distorts competition notably in terms of prices (similar discourses were raised in other countries – see for example Horn and Merante 2017). During our research, no particular public (including media) and social partner attention was detected on the labour issues pertaining to the operation of the big Airbnb ecosystem in Greece. Thus, the fact that a considerable number of people seemed to work in the rental economy of Airbnb under

conditions of precarious employment (Kathimerini 2018a) and undeclared work went largely unnoticed (on these issues, see Schor 2014).

Rather, tax evasion and the provision of services by unlicensed operators was the primary aspect tackled by policymakers and regulators – an
aspect also largely raised by hotel owners. As argued by an interviewed
representative of a social partner organisation: "Neither the social dialogue includes Airbnb's phenomenon nor the new law reforms about
sharing economy's accommodation sector, since it mainly regulates the
taxes, without touching the issues of labor or home and neighborhood
protection." The growth of Airbnb accommodation led to Greek hotels
losing 12 million overnight stays, which translated into €554 million
less in revenue, accounting for 15,000 job losses per year from the hotel
sector and a reduction in taxes of up to €350 million per year (Greek
Travel Pages 2015). As representatives from social partner organisations
in the hotel industry said, the sharing economy need no longer be considered as an alternative minor activity, but rather as a large economic activity generating high turnover and working in parallel with the licensed
sector (ibid.). In seeking to address this development, from early June
2016, the Greek Tourism Ministry, following similar legislation imposed
in Germany, introduced a cash tax on Airbnb-style rentals and it also
threatened to fine those owners who did not register the properties they
advertised on Airbnb as businesses.

As for second-level or legislative changes, in 2016, a regulative
framework regarding short-term leases through online platforms was
introduced by virtue of Article 111 of Law 4446/2016 (Government
Gazette Bulletin A' 240/22.12.2016), which is titled, "Arrangements for
short-term rental of properties in the context of the sharing economy"
(our translation).[5] Greece thus became one of the first countries to regulate short-term home rentals (Povich 2014). Article 111 provided the
following important definitions:

– "A *sharing economy*" is considered to be any model where digital
  platforms create an open market for the temporary use of goods or
  services that are often provided by individuals.
– "*Digital platforms*" are electronic, bilateral or multilateral markets
  where two or more user groups communicate via internet with the
  mediation of a platform manager facilitating a transaction between
  them.

Specific provisions were amended and enhanced by virtue of Articles 83 and 84 of Law 4472/2017 (Government Gazette Bulletin A' 74/19.05.2017). Parliament ratified the bill after Airbnb refused to turn over the personal data of property owners registered with the online platform and customers. The new law requires that property owners who use digital platforms to lease accommodation to tourists for short periods pay up to 45 per cent in tax on their income, with the purpose of boosting state funds.[6] Though some analysts have warned that the measure might end up discouraging potential visitors from visiting Greece, representatives of social partner organisations in the real estate sector supported the new law, arguing that it can potentially bring "order and balance in a situation which until recently was totally anarchic" (Kathimerini 2018b). However, grievances were expressed concerning the absence of a social dialogue framework to guide the legislative changes brought forward; as one of our interviewees (a representative of a Greek tourist organisation) stressed, "Social dialogue has not been used effectively overall in the process of enacting reforms in the sharing economy's accommodation sector, since only a small part of the social partners which are involved with Airbnb actually collaborated with the government in devising these reforms", a point also shared by an interviewed trade unionist in the sector of tourism, who remarked, "shadow economy-activities in the platform-type business have created in Greece an unfair competition context."

2.2 Policy and Legislative Changes: The Case of Uber

In the transport sector, Uber gave rise to much media attention. Uber in Greece was launched in Athens in 2014 and in the following years it ran two services: UberX with private drivers and UberTAXI with professional taxi drivers. In April 2018, Uber announced that it would suspend its licensed service UberX (with private drivers) after the approval of new legislation that imposes stricter regulation on Uber and Uber-like platforms.

The new Law 4530/2018, titled "Arrangements for transportation issues and other provisions" (our translation), foresees among others that e-platforms and apps offering taxi services as intermediaries operate as transport companies, requiring them to enter three-year contracts with licensed taxi drivers and outlining the "exact terms of use of the brokerage service as applicable, the data of owners or drivers of the vehicles, vehicle registration certificates and special driving licenses for vehicle drivers." In effect, the bill breaks away from past provisions that aimed to liberate the taxi market and allowed, among others, tourist agencies to lease cars and drivers at will.

The new law foresees harsh fines. Should taxi service providers operate without a licence issued by the Transport Ministry, they will be fined up to €100,000 and the driver will have his professional license revoked for two years. At the same time, vehicle owners "cooperating illegally with a mediating taxi service will be subject to an administrative fine of 5,000 Euros per vehicle." The new law obliges companies active in the sector to move forward with issuing a licence if the electronic or telephone brokerage service is their main activity. This legislative move, initially aimed at hindering Uber's market presence, also impacted a great number of taxi service providers collaborating with individual drivers. One such company is the popular ride-hailing app Beat (launched in 2011 by a Greek entrepreneur). The service provided by Beat in Athens is similar to UberTAXI, which Uber still maintains. In particular, once governmental intentions became known, the founder of Beat launched an online campaign to stop the government from drafting the new bill, with over 30,000 supporters signing the online petition (Greek Travel Pages 2017b).

It should also be stressed that the new law was preceded by angry demonstrations by traditional taxi drivers who took to the streets of Athens to protest against Uber and Beat for taking away their business (some even attacked a vehicle they believed to be driven by an Uber worker). Several media reports of the time suggested the government acted under pressure from traditional taxi drivers and their powerful and influential organisations, while, once the new law was passed, Greece's major political parties in opposition accused the Transport Ministry of hampering innovation and entrepreneurship. In response, Transport Ministry officials underlined that the bill was in line with EU legislation and a recent European Court ruling placing emphasis on the need to address potential tax evasion and avoidance practices on the part of digital platforms (Kathimerini 2018c).

As with Airbnb, and contrary to what has been happening in other European countries, no particular attention (policy or otherwise) was given to issues of employment rights, working conditions and social protection faced by Uber drivers. The observed (quasi-)non-contention around critical labour questions pertaining to gig workers in Greece and how to regulate them – but also on the potential of new job opportunities that could be generated with the growth of the gig economy (a considerable aspect for a country with one of the highest unemployment rates in Europe) – is difficult to decipher unless one looks more closely at the historical roots and evolution of the Greek model of social partnership and trade unionism; and more particularly at the persistence of strong legacies that seem to have remained salient even after the emergence of digital platforms in Greece.

2.3 Responses of Workers and Employers

Second-level changes also encompass the degree of change of social partners' mobilisations and representation tactics and strategies. The spread of flexible, non-standard work in the gig economy raises concerns about respecting freedom of association, workers' representation and collective bargaining. In the case of social partners, the international literature touches on the need for workers' and employers' organisations to anticipate and adapt their organising tactics and collective bargaining approaches to the multifaceted demands of digital gig economy and work (Vandaele 2018). Three main approaches to organising gig workers are identified[7] – that is, the legal or judicial approach, which mainly entails unions contesting worker misclassification as self-employed; alliance formation aimed at providing support services to gig workers; and efforts for advocacy and lobbying in order to exert pressure towards the introduction of new local or national legislation (see Introduction in this book).

Looking more closely at the Greek case, we find no concrete initiatives falling in any of the three approaches described above. At the time of writing, no major organising initiatives have been taken (that is, initiatives to organise digital gig workers) by trade unions to influence workplace politics and conditions in the gig economy; and no major knowledge brokering activities[8] have been initiated. Furthermore, no specific regulatory initiatives or collective agreements have been undertaken by unions to specifically address the gig economy and its effects on labour and social protection rules. Overall, Greek unions did not seek to engage with groups of gig workers as part of a renewal strategy to expand representation to non-standard workers (more broadly), as international evidence shows in relation to other countries.[9]

A more active approach in terms of knowledge brokering was taken by the big confederation of Greek employers[10] and the Economic and Social Council of Greece.[11] The most critical voices, as discussed above, were organisations representing businesses that were most directly affected by digital platforms in the transport and the hospitality and tourism industry (for example, traditional taxi drivers, tourist accommodation enterprises and hoteliers). Through repeated strikes and public criticism, they were able to increase political pressure for regulatory reforms. We could interpret their activist approach as falling within the third approach identified above – that of pushing for regulatory reforms. Interestingly, these cases concern responses taken by employer (rather than worker) organisations, and they are explicitly aimed at curtailing what are generally perceived as negative effects of other (but still) employers in the gig economy. This corroborates the argument advanced by Kirchner and Beyer (2016) according to which the rise of the broader platform economy has been followed by competition between different logics or models within capitalism – that is, between employers in platform capitalism and employers in conventional or mainstream capitalism operating in the same industry or sector.

Due to emerging intra-capitalist conflict in which, as Rinne (2018) writes, access (to assets) over ownership (of assets) became a new tagline, the balance of power between capital and labour found in regular employment has also been disrupted. Thompson established the Disconnected Capitalism Thesis in the early 2000s to refer to the rising gap "between what capital is seeking from employees ... and what it finds necessary to enforce in the realm of employment relations [employment relationships]" (Thompson 2003, p. 264). According to this viewpoint, firms have prioritised downsizing and divestment strategies in response to financialisation[12] and flexibilisation[13] (Rubery et al. 2016), and they have withdrawn from investment in human capital, with consequences in terms of career progression, employment protection and pensions. A large body of research (Grimshaw et al. 2016; Fitzgerald et al. 2012; Kalleberg 2000, 2011; Milkman and Ott 2014; Pollert and Charlwood 2009) has examined how such practices undermine employees' ability to alter their employment situation.[14] The practices of ride-hailing platform companies, which rely on paying work that is contracted on short notice (that is, paying drivers by the ride) and on the basis of existing demand (only if there are rides) thereby on significant income volatility (Farrell and Greig 2016), best exemplify this trend of work flexibilisation, which denotes a reformulation of the very notion of "work" itself (Grossman and Woyke 2016; Kenney and Zysman 2015) as well as a shift away from the paradigmatic form of standard employment defined as a job with a fixed number of hours and for a predetermined salary and social protections (De Stefano 2016; Drahokoupil and Fabo 2016; Felstiner

2011). Many of the regulatory and social welfare achievements of the 20th century – minimum pay, paid sick leave, social insurance coverage and protection from unfair dismissal – and safeguards of "decent work" (ILO, 2019) have not been sufficiently expanded to include gig work (Greenhouse 2015; Mexi 2020; OECD 2019). Even in extreme situations such as the Covid-19 pandemic crisis, according to a survey on workers on digital platforms in the taxi and delivery sectors: "Seven out of ten workers indicated not being able to take paid sick leave, or to receive compensation, in the event they were to test positive for the virus, thus risking the health of others in addition to their own health" (ILO 2021, p. 25).[15] Unbalanced power relations between digital platforms and gig workers are thus becoming more visible, necessitating a rethinking of how policy and regulation might address both the effects and drivers of such fundamental imbalances (Grosheide and Barenberg 2016; Harris and Krueger 2015; ILO 2019; JRC 2016; Kennedy 2016; Lobel 2016; Malhotra and Van Alstyne 2014; Mexi 2021; Prassl and Risak 2016; Ranchordas 2015; Rauch and Schleicher 2015).

Although the Greek case shows no intense conflict between gig workers and platforms as is the case, for example, with Uber and Uber taxi drivers in the United Kingdom, there is evidence that the spread of Airbnbing has provided certain groups with a new impetus for organising themselves for collective representation and discussions. Hence, Airbnbing has led to the emergence of a new collective organisation in Greece: the Association of Sharing Economy, S.O.DIA. Founded in 2017 in order to represent owners, sub-tenants and third-party managers of Airbnb-type of accommodation, S.O.DIA. participated in the discussions with the Greek Independent Authority for Public Revenue (AADE) preceding the introduction of the regulatory initiatives taken in 2017. In their public announcements at that time, S.O.DIA seemed to align its efforts with AADE on the need to protect the market from monopolies, stressing also, "The sharing economy has a positive contribution to the economy as it promotes the concept of Greek hospitality, strengthens local products and businesses and makes use of assets that otherwise would remain inactive" (Greek Travel Pages 2017a).

Beside the emergence of S.O.DIA, so far we have found no evidence of grassroots initiatives taken by people working in the gig economy to achieve effective representation and to build collective agency. In particular, we have discerned no effort for unionising Greek gig workers, as can be observed elsewhere (Johnston and Land-Kazlauskas 2019; Roberts 2018). Yet, all this is happening against a background of reported dissatisfaction about unstable working hours, physical exhaustion and time pressure on the part of people working in the wider Airbnb business.[16]

3.   ASSESSING THE GREEK CASE: PAST PATHOLOGIES AND NEW LIMITATIONS

To capture the salience of variables or factors mediating Greek unions' and employers' low-to-moderate responses (vis-à-vis the social partners' more active responses in other European countries) to the new realities of gig work, a more in-depth understanding of certain national peculiarities pertaining to the Greek model of social partnership is needed, as explained in more detail below. In particular, three key issues were mentioned during the interviews.

3.1   The Greek Model of Social Partnership and its Connection to the State Prior to the Rise of the Gig Economy

A historical look at the nature and evolution of the Greek model of social partnership helps to understand the difficulty of adapting to the concept and realities of the gig economy and work, while also hinting at potential solutions. Excessive reliance on party-political influences, as well as historical legacies of clientelism and political patronage (Mouzelis 1986) and networks of comprehensive internally fragmented interest structures (see below, 3.3), have contributed to the formation of "disjointed corporatism" (Lavdas 1997, 2005) as the main form of interest representation. This is characterised by a highly fragmented type of interest mediation in which the labour movement remains internally split among its peak organisations, with trade unions largely acting as "rent-seeking" players (Voskeritsian et al. 2019). These specific characteristics can also be associated with the social dialogue regression over the crisis years (ILO 2013) and particularly with the observed weakness of mainstream trade unions to adequately address the austerity challenge and its adverse implications on workers (Greer and Doellgast 2013). Since 2010, there has been substantial popular mobilization against the austerity measures included in the bailout package presented to Greek governments by the Troika of creditors (EU–ECB–IMF). Nonetheless, institutional trade unions have failed to halt the decline in wage earners' income, which had plummeted by half by 2013 compared with 2008. These unions have also generally failed to establish their leadership position in rallying the working population. As Kretsos and Vogiatzoglou (2015, p.223) write, "although trade unions challenged sometimes significant efforts of marketisation in the pre-crisis period through strike action (e.g. proposing social security reforms in 2003), they remained among the main stakeholders of the mainstream political and institutional order in Greece, reflecting the public discourse and perception of trade unions as 'political dinosaurs'."

3.2   Trade Unions' "Business as Usual" and the Lack of a Social Dialogue Framework to Address Emerging Problems

Trade unions continue to operate in a more traditional framework when it comes to labour issues associated with the gig economy. During the timeframe of our research, we did not detect any evidence of trade unions opening up to workers in the gig economy. Overall, the result of that has been, as one interviewee working in the Airbnb business said, "an emerging group of gig workers most of them working informally in the booming Airbnb business that are left alone, vulnerable and unprotected." This "inclusiveness deficit" can be interpreted through different perspectives. To begin with, trade union (quasi-absence of) responses vis-à-vis gig workers may be understood against their limited interest and capacity to adequately reach out to precarious workers, as past experience shows. *Mutatis mutandis*, the case of migrants and young workers bears certain similarities with gig workers, a point brought forward by an interviewed journalist. As argued, migrant workers from Africa and Asia were overrepresented during the 1990s and mid-2000s in the booming industries of construction and agriculture. Nevertheless, this new reserve army of labour remained without union representation, which resulted in the development of direct employer control leading to exploitation at work.

Another explanation may relate to the perceived costly allocation of resources to address the needs of gig workers. That is to say that the low numbers of gig workers in the general Greek workforce (though reliable figures are lacking) seem to explain the low interest on the part of the trade unions to commit resources to gig workers' organisation and protection. According to an interviewed trade unionist active at national level, "This [the situation of gig workers] is a tiny problem compared to all the other issues that Greek employees face." Along similar lines, another interviewed trade unionist in the tourist sector said, "[S]haring economy's accommodation sector employees must be few in number, not declared to public authorities and difficult to be identified."

Moreover, trade unions' lack of interest in mobilising and protecting the precarious workforce in the gig economy may also be seen as part of a wider disillusionment that emerged after the onset of the economic crisis and the 2010 bailout agreement with Greece's creditors. As one interviewed expert stressed, "Especially during the early crisis years, the mainstream trade unions were not sufficiently able to impede the reduction of wage earners' income while the overall resilience of social dialogue came under pressure."

Lastly, an interviewed expert critically assessing the role of social partners more broadly referred to the lack of an efficient social dialogue framework to address the issues at stake: "New topics are not discussed or debated in a context of social dialogue, although they should in order to provide a way out of the economic crisis for young people and other vulnerable people.... There is also the issue of power resources of the unprotected flexible labour force in the gig economy in Greece, or rather, their lack of them." The latter point resonates with Rainnie and Ellem's argument (2006) that the reorientation of trade union identities, approaches and strategies is strongly related to realities and changes in the political balance of power in Greece. This observation is essential for a country with a heavily politicised social partnership environment.

3.3   Fragmentation and the Traditional Absence of a Genuine and Intense Consensus-Seeking Culture

Another factor behind the low-level responses of social partner organisations in Greece in terms of addressing the contentious labour issues associated with the gig economy may be related to their traditionally fragmented nature. A plethora of associations and groups has divided employers and employees into interest groups lacking coherence and strong organisational resources, preventing in several instances their effective emancipation from state structures and clientelistic politics. Reforms to institutionalise collective bargaining and structures conducive to social dialogue in the 1980s and 1990s may have contributed to facilitating ad hoc instances of consensus and concerted action (Daskalakis 2015). However, the effects of fragmentation on policymaking have persisted and become more complicated when one considers that Greek political culture – under which the evolution of social partnership is subsumed – has long been characterised by party polarisation exacerbated by the civil war during the 20th century (Andreadis and Stavrakakis 2019) and the lack of conciliatory mechanisms to facilitate substantial policy reform (Sotiropoulos 2012, 2017).

Currently a broad-based consensus on a joint strategy for a socially inclusive economic growth approach to preparing for and managing the gig economy is lacking. Such an approach could be based on a consensual rapprochement between the social partners and the gig economy actors. It could also outline a range of policy priorities to profit from both business innovation and the employment opportunities offered through the gig economy, while seeking to combat the problems of precarity and inequality in the labour market. And it could put forward specific reforms to reorganise work and welfare towards a more sustainable growth model, as Greece is entering its post-crisis era. International experience shows that social dialogue could play a key role in this respect (Mexi 2019). As characteristically stressed by one of our interviewees (representative of an organisation active in the tourist and real estate sectors), "The scope of social dialogue concerning the gig economy is limited at the moment, but needs to be strengthened, because new legislative reforms should definitely be implemented in the future. Collective bargaining and social dialogue will gain prominence only through the wide consultation of all social partners, but this will require time and planning."

4.   CONCLUSION

Overall, our study corroborates the argument in the literature that digital labour platforms are not only a matter of technological innovation and change. Context matters. The advance of gig economy and work should be contextualised within distinct societies. Responses to gig economy developments in Greece are shaped to a considerable extent by the historical legacies of the social and political framework in place. In particular, our findings indicate that, as the country grappled with economic crisis and disillusionment, past legacies have persisted and they constitute one key factor hindering the introduction of major reforms to address both job growth and innovation opportunities as well as the labour challenges pertaining to the gig economy and work.

More particularly, while there is evidence of low-to-moderate first-level (ideational/cognitive) change, no policy-oriented changes at the second level aiming to address questions related to gig work and how to regulate it could be found. During the timeframe of our research, we detected no significant evidence of social partners' role in the formulation and implementation of reforms to solve specific labour problems, those resulting for instance from the rise of flexible, precarious, casual and non-standard employment (especially with regard to the platforms operating in Greece's expanding tourism sector). To date, there have been no attempts to formulate national plans or strategies to maximise opportunities and reduce risks. The fact that legal interventions and discourses primarily addressed tax-related issues and not the more controversial and demanding labour-related issues – concerning, for instance, the status of gig workers or their working conditions and the quality of jobs offered via digital platforms – points to the presence of historical factors that are related to the country's specific institutional and political conditions and the Greek model of social partnership and trade unionism. It can be argued that these factors have traditionally contributed to blocking the re-articulation of policy preferences along more comprehensive transformative policies that could eventually be used to address the needs of groups who are lacking any mobilisation resources, effective rights or the ability to exercise more political pressure.[17]

Summing up, in our study we have sought to show that the social partners' (lack of) ability to make the most of the gig economy can be explained by the salience of mediating contextual variables and path dependencies. Yet, despite the persistence of past legacies, possibilities to discontinue from old ways of doing things do exist for Greek social partners as long as they come to realise the significance of strengthening their organisational abilities and making social dialogue more inclusive and representative as well as enhancing their policy responses as a result. As the literature on historical institutionalism (Beckert 1999; Garud and Karnøe 2001) indicates, the rise of "institutional entrepreneurs" defined as "individuals whose creative acts have transformative effects on politics, policies, or institutions" (Sheingate 2003, p.185) has the potential to interrupt legacies and change dynamics, thereby transcending discontinuities and adverse institutional histories on the ground. Currently, and in anticipation of path-breaking actors, only hindsight can suggest whether the gig economy will provide for Greece an entirely revolutionised trajectory of economic growth or instead an incremental evolution of economic activity that becomes absorbed in prevailing social partnership legacies without any paradigmatic change. In a country emerging from the crisis with a highly traumatised economy and society, policymakers, employers and workers need to understand the importance of leveraging the opportunities of the gig economy into innovation combined with socially inclusive growth. Only in this way, they will not allow continuities with the past becoming the wrong blueprints of the future.

NOTES

1. Eurostat, http://appsso.eurostat.ec.europa.eu/nui/show.do?dataset=nama_10
   _gdp&lang=en (accessed on 3 October 2022).
2. Eurostat, http://appsso.eurostat.ec.europa.eu/nui/show.do?dataset=une_rt_a
   &lang=en (accessed on 3 October 2022).
3. Some of these platforms are likely to be co-host or management companies
   that manage Airbnb listing for hosts, either through the Airbnb platform
   itself (co-hosts) or outside of it.
4. At the same time, Airbnb established its Office of Healthy Tourism with
   the mandate to address issues of overtourism in cities and to "foster initiatives that drive economic growth in communities, empower destinations
   from major cities to emerging destinations, and support environmental
   sustainability" (see: https://www.airbnbcitizen.com/officeofhealthytourism/
   , accessed on 3 October 2022).
5. The law sets a series of strict conditions that must be met cumulatively in
   order for property rental to be lawful. These conditions prevent the lessor
   from acquiring a commercial capacity through the provision of organised
   accommodation services (Felouka 2018).
6. In the legislation, the Finance Ministry stipulates categories of tax rates
   that will be borne by the property owners profiting from the tourism platforms' sharing economy. Specifically, owners receiving up to €12,000 in
   revenues will be taxed at 15 per cent, while others receiving from €12,001
   to €35,000 and amounts greater than €35,001 will be taxed at 35 per cent
   and 45 per cent, respectively. Airbnb-style hosts, defined as "operators",
   will be required to enter the registry of the Independent Authority for Public
   Revenue, submit a short-term residence declaration for each tenant, enroll
   in the short-term residential property data system, inform the Deposits and
   Loans Fund for income attributable to unknown beneficiaries, as well as
   provide information on tenants and duration of stay. The decision underlined
   that there can only be one "operator" per property. The law, which foresees
   fines of up to €5,000 for violators, was also signed by the head of Greece's
   Independent Authority for Public Revenue (Felouka 2018).
7. These vary across geographies, depending on the political context in which
   they operate.
8. Activities such as training, consultancy, conferences and networking.
9. For a comprehensive examination of the issues of collective voice and
   organising gig workers, see particularly Johnston and Land-Kazlauskas
   (2019).
10. The Hellenic Federation of Enterprises, SEV (the biggest organisation repre-
    senting employers in Greece), organised a conference and published a report
    (2017) on the future of work within the wider context of digitalisation
    of society and the economy. The report sheds light on the confusion sur-
    rounding several related concepts, namely, the "digital economy", "sharing
    economy", "collaborative economy", "gig economy", "crowdwork" and
    "gig work", aiming to bring some clarity and suggest first-level ideational
    changes among its members. The report also stresses the need for more

targeted awareness-raising for employers to harness the impending effects of the digital revolution on the labour market and the economy.

11. In this regard, worth mentioning is the Conference on "Social Dialogue and the Future of Work" organised jointly by the Economic and Social Council of Greece, OKE (composed of representatives of Greece's major social partner organisations); the International Labour Organization, ILO; and the International Association of Economic and Social Councils and Similar Institutions in Athens in November 2017. The aim of the conference was to trigger a more dynamic and continuous dialogue between government, employer and worker representatives on the transformational challenges related to the digitalisation of societies and economies, and to promote the sharing of good practices. The debates around the disruptiveness of the gig or platform economy were identified as key issues of concern along with the need for more active participation of social partners in the design and decision-making affecting the future of work.

12. The process of financialisation refers to profits increasingly generated through financial channels and investments rather than production activities or productive value-added services as financial deregulation has enabled more volatile investments (Findlay et al. 2017).

13. This concept gained momentum globally already in the 1980s, during periods of severe unemployment and intense competition (Benassi 2013; Keune and Serrano 2014; Grimshaw et al. 2016; Rubery et al. 2016).

14. According to Thompson (2003), financialisation processes have involved significant implications for work and employment relationship dynamics. In particular, as Lazonick and O'Sullivan (2000) write, the demand for short-term financial results requires flexibility with the possibility that workers are laid off. It also involves a preference for individualistic performance-related pay systems (made easier with the algorithmic systems of the platforms in the gig economy), rather than firm-specific skills, and hostility towards union bargaining (Jacoby, 2005).

15. According to Grimshaw et al. (2016), the existence of specific protective gaps in the areas of social protection, regulation and representation exacerbate non-standard and flexible employment arrangements, namely lack of access to pensions and unemployment protection, lack of resources to cover employment tribunal fees, and lack of access to union membership.

16. See, for example, the article *Airbnb: «Είναι σαν μία κανονική δεύτερη δουλειά χωρίς ωράριο»* published in the newspaper *Kathimerini* (2018a), which presents the true stories of Greek young persons (mostly graduates) who have been working in the wider Airbnb business in Athens either as co-hosts or as super-hosts (undertaking several tasks such as managing bookings, connecting with guests, contracting cleaners, etc.), or as drivers (collaborating with hotels and car owners renting their cars via apps). While describing their work as "a real full-time job without fixed working hours", they have also been complaining about physical fatigue due to unpredictable scheduling, inconsistent earnings, unreliable long-term employment prospects, the last preventing them from making major life decisions, such as starting a family.

17. In this regard the "reform technology" (Monastiriotis and Antoniades 2009) of the country to address the emerging gig economy developments has proven weak.

REFERENCES

Academic Network for European Disability Experts – ANED (2015/2016), "European Semester 2015/2016 Country Fiche on Disability", available at: http://www.disability-europe.net/ (accessed on 3 October 2022).

Andreadis, I., and Y. Stavrakakis (2019), "Dynamics of Polarization in the Greek Case", *The ANNALS of the American Academy of Political and Social Science*, **681**(1), 157–167. https://journals.sagepub.com/doi/10.1177/0002716218817723.

Badcock, J. (2017), "Barcelona Threatens to Expel Airbnb from the City after Accusing Executive of Illegally Listing Sublet", *The Telegraph*, available at: https://www.telegraph.co.uk/news/2017/06/27/barcelona-threatens-expel-airbnb-city-accusingexecutive-illegally/ (accessed on 3 October 2022).

Beckert, J. (1999), "Agency, Entrepreneurs, and Institutional Change: The Role of Strategic Choice and Institutionalized Practices in Organizations", *Organization Studies*, **20**(5), 777–799.

Benassi, C. (2013), "Political Economy of Labour Market Segmentation: Agency Work in the Automotive Industry", ETUI Working Paper.

Daskalakis, D. (2015), *Greek Labour Relations in Transition in a Global Context*, Peter Lang GmbH.

De Stefano, V. (2016), "The Rise of the 'Just-in-Time Workforce': On-Demand Work, Crowdwork and Labour Protection in the 'Gig-economy'", *Comparative Labor Law Policy Journal*, **37**(3), 471–504.

Drahokoupil, J., and B. Fabo (2016), "The Platform Economy and the Disruption of the Employment Relationship", ETUI Policy Brief No 5/2016.

Farrell, D., and F. Greig (2016), "Paychecks, Paydays, and the Online Platform Economy Big Data on Income Volatility", Annual Conference on Taxation and Minutes of the Annual Meeting of the National Tax Association.

Felouka, D. (2018), "AIRBNB – Legislation and Tax Transparency", available at: http://www.greeklawdigest.gr/topics/real-estate/item/288-airbnb-legislation-and-tax-transparency (accessed on 19 September 2022).

Felstiner, A. (2011), "Working the Crowd: Employment and Labor Law in the Crowdsourcing Industry", *Berkeley Journal of Employment and Labor Law*, **32**(1), 143–204.

Findlay, P., P. Thompson, C. Cooper and R. Pascoedeslauriers (2017), *Creating and Capturing Value at Work: Who Benefits?*, London: Chartered Institute of Personnel and Development.

Fitzgerald, I., J. Hardy and M. Martinez Lucio (2012), "The Internet, Employment and Polish Migrant Workers: Communication, Activism and Competition in the New Organisational Spaces", *New Technology, Work and Employment*, **27**(2), 93–105.

Garud, R., and P. Karnøe (2001), "Path Creation as a Process of Mindful Deviation", in R. Garud and P. Karnøe (eds), *Path Dependence and Creation*, Mahwah, NJ: Lawrence Erlbaum, pp. 1–38.

Goranitis, Y. (2016), "Airbnb: Το Οικονομικό Θαύμα της Διπλανής Πόρτας ή μια Παγκόσμια Φούσκα", *The Insidestory*, available at: https://insidestory.gr/article/airbnb?token=QHAIUPIS6P (accessed on 3 October 2022).

Greek Travel Pages (2015), "The 'Sharing Economy' in Tourism Deprives Greece of Millions of Euros", October, available at: https://news.gtp.gr/2015/10/08/sharing-economy-tourism-deprives-greece-millions/ (accessed on 3 October 2022).

Greek Travel Pages (2017a), "Airbnb Hosts in Greece Will Have to Register Their Property", available at: https://news.gtp.gr/2017/11/22/airbnb-hosts-greece-register-property/ (accessed on 3 October 2022).

Greek Travel Pages (2017b), "Draft Bill Prompts Greece Taxi Service to Launch Online Petition", available at: https://news.gtp.gr/2017/09/29/draft-bill-prompts-greece-taxi-service-launch-online-petition/ (accessed on 3 October 2022).

Greenhouse, S. (2015), "Uber: On the Road to Nowhere", *The American Prospect*, available at: http://prospect.org/article/road-nowhere-3 (accessed on 3 October 2022).

Greer, I., and V. Doellgast (2013), "Marketization, Inequality, and Institutional Change", available at: http://gala.gre.ac.uk/10294/ (accessed on 3 October 2022).

Grimshaw, D., M. Johnson, J. Rubery and A. Keizer (2016), "Reducing Precarious Work in Europe through Social Dialogue: Protective Gaps and the Role of Social Dialogue in Europe", Report for the European Commission, Institute of Work, Skills and Training, University of Duisburg, Essen.

Grosheide, E., and M. Barenberg (2016), "Minimum Fees for the Self-Employed: A European Response to the 'Uber-ized' Economy?", *Columbia Journal of European Law*, **22**(2), 193–236.

Grossman, N., and E. Woyke (2016), *Serving Workers in the Gig Economy Emerging Resources for the On-Demand Workforce*, Sebastopol, CA: O'Reilly Media.

Gutiérrez, J., J.C. García-Palomares, G. Romanillos and M.H. Salas-Olmedo (2017), "The Eruption of Airbnb in Tourist Cities: Comparing Spatial Patterns of Hotels and Peer-to-Peer Accommodation in Barcelona", *Tourism Management*, **62**, 278–291.

Guttentag, D. (2013), "Airbnb: Disruptive Innovation and the Rise of an Informal Tourism Accommodation Sector", *Current Issues in Tourism*, **18**(12), 1192–1217.

Hall, P.A. (1993), "Policy Paradigms, Social Learning, and the State: The Case of Economic Policymaking in Britain", *Comparative Politics*, **25**(3), 275–296.

Harris, S., and A. Krueger (2015), "A Proposal for Modernizing Labor Laws for Twenty-First-Century Work: The 'Independent Worker'", Discussion Paper 2015-10, The Hamilton Project, Brookings, Washington, D.C.

Horn, K., and M. Merante (2017), "Is Home Sharing Driving Up Rents? Evidence from Airbnb in Boston", *Journal of Housing Economics*, **38**, 14–24.

ILO (2013), "Social Dialogue: Recurrent Discussion under the ILO Declaration on Social Justice for a Fair Globalization", Report VI, International Labour Conference 102nd Session, ILO, Geneva.

ILO (2019), *Work for a Brighter Future – Global Commission on the Future of Work*, Geneva: ILO.

ILO (2021), *World Employment and Social Outlook: The Role of Digital Labour Platforms in Transforming the World of Work*, Geneva: ILO.

Jacoby, S. (2005), *The Embedded Corporation: Corporate Governance and Employment Relations in Japan and the United States*, Princeton, NJ: Princeton University Press.

Johnston, H., and C. Land-Kazlauskas (2019), "Organizing On-Demand: Representation, Voice, and Collective Bargaining in the Gig Economy", Inclusive Labour Markets, Labour Relations and Working Conditions Branch, ILO, Geneva.

JRC (2016), *The Future of Work in the "Sharing Economy", Market Efficiency and Equitable Opportunities or Unfair Precarisation?*, JRC Science for Policy Report.

Kalleberg, A. L. (2000), "Non-Standard Employment Relations: Part-Time, Temporary and Contract Work", *Annual Review of Sociology*, **26**(1), 341–365.

Kalleberg, A. L. (2011), *Good Jobs, Bad Jobs: The Rise of Polarized and Precarious Employment Systems in the United States, 1970s to 2000s*, New York: Russell Sage Foundation.

Kathimerini (2018a), "Airbnb: «Είναι σαν Μία Κανονική Δεύτερη Δουλειά Χωρίς Ωράριο»", available at: http://www.kathimerini.gr/961631/article/oikonomia/real-estate/airbnb-einai-san-mia-kanonikh-deyterh-doyleia-xwris-wrario (accessed on 3 October 2022).

Kathimerini (2018b), "New Tax Legislation Likely to Shake Up Short-term Property Lease Market", available at: http://www.ekathimerini.com/224922/article/ekathimerini/business/new-tax-legislation-likely-to-shake-up-short-term-property-lease-market (accessed on 3 October 2022).

Kathimerini (2018c), "Ride-hailing App Uber Halts its Licensed Service in Greece", available at: http://www.ekathimerini.com/227452/article/ekathimerini/business/ride-hailing-app-uber-halts-its-licensed-service-in-greece (accessed on 3 October 2022).

Kennedy, J. (2016), "Three Paths to Update Labor Law for the Gig Economy", ITIF, available at: https://itif.org/publications/2016/04/18/three-paths-update-labor-law-gig-economy (accessed on 3 October 2022).

Kenney, M., and J. Zysman (2015), "Choosing a Future in the Platform Economy: The Implications and Consequences of Digital Platforms", Discussion paper, Kauffman Foundation New Entrepreneurial Growth Conference, Amelia Island Florida, 18 June.

Keune, M., and A. Serrano (eds) (2014), *Deconstructing Flexicurity and Developing Alternative Approaches: Towards New Concepts and Approaches for Employment and Social Policy*, Abingdon: Routledge.

Kirchner, S., and J. Beyer (2016), "Die Plattformlogik als digitale Marktordnung. Wie die Digitalisierung Kopplungen von Unternehmen löst und Märkte transformiert", *Zeitschrift für Soziologie*, **45**(5), 324–339.

Kretsos, L., and M. Vogiatzoglou (2015), "Lost in the Ocean of Deregulation? The Greek Labour Movement in a Time of Crisis", *Le syndicalisme en quête d'autonomie et de renouvellement en Europe*, **70**(2), 218–239.

Lavdas, K. (1997), *The Europeanization of Greece: Interest Politics and the Crises of Integration*, London: Macmillan.

Lavdas, K. (2005), "Interest Groups in Disjointed Corporatism: Social Dialogue in Greece and European 'Competitive Corporatism'", *West European Politics*, **28**(2), 297–316.

Lazonick, W., and M. O'Sullivan (2000), "Maximising Shareholder Value: A New Ideology for Corporate Governance", *Economy and Society*, **29**(1), 13–35.

Lemonis, V. (2015), "Airbnb, Sweet Airbnb: Hosts' Perspectives on Managing Commercial Homes and Offering Experiences", 1st International Conference on Experiential Tourism, available at: http://scholar.googleusercontent.com/scholar?q=cache:5ms-IPIqNugJ:scholar.google.com/+Airbnb+greece&hl=en&as_sdt=0,22 (accessed on 3 October 2022).

Lobel, O. (2016), "The Law of the Platform", *Minnesota Law Review*, **101**(1), 88–166.

Malhotra, A., and M. Van Alstyne (2014), "The Dark Side of the Sharing Economy… and How to Lighten It", *Communications of the ACM*, **57**(11), 24–27.

Mexi, M. (2021), "The Platform Economy – Time for More Democracy at Work", Social Europe, available at: https://socialeurope.eu/the-platform-economy-time-for-more-democracy-at-work (accessed on 3 October 2022).

Mexi, M. (2020), "The Future of Work in the Post-Covid-19 Digital Era", Social Europe, available at: https://socialeurope.eu/the-future-of-work-in-the-post-covid-19-digital-era (accessed on 3 October 2022).

Mexi, M. (2019), "Social Dialogue and the Governance of the Digital Platform Economy: ILO and the Role of Social Partners", Paper prepared within the framework of the SNIS-funded project *Gig Economy and its Implications for Social Dialogue and Workers' Protection*, ILO, Geneva.

Mexi, M. (2018), "Greece in Times of Multiple Crises: Solidarity under Stress?", in V. Federico and C. Lahusen (eds), *Solidarity as a Public Virtue? Law and Public Policies in the European Union*, Baden-Baden, Germany: Nomos Verlag, pp. 337–60.

Milkman, R., and E. Ott (eds) (2014), *New Labor in New York: Precarious Workers and the Future of the Labor Movement*, Ithaca, NY: ILR Press.

Monastiriotis, V., and A. Antoniades (2009), *Reform That! Greece's Failing Reform Technology: Beyond "Vested Interests" and "Political Exchange"*, London: The Hellenic Observatory, London School of Economics and Political Science, GreeSE paper 28.

Mouzelis, N. P. (1986), *Politics in the Semi-periphery: Early Parliamentarism and Late Industrialization in the Balkans and Latin America*, London: Macmillan.

OECD (2019), "OECD Employment Outlook 2019: The Future of Work", OECD, available at: https://www.oecd-ilibrary.org/sites/b40da5b7-en/index.html?itemId=/content/component/b40da5b7-en (accessed on 3 October 2022).

Pollert, A., and A. Charlwood (2009), "The Vulnerable Worker in Britain and Problems at Work", *Work, Employment and Society*, **23**(2), 343–362.
Poutetsi, Ch. (2018), "Η Airbnb Ανακοινώνει Στοιχεία και για Ελλάδα και Παίρνει Θέση για τον Υπερτουρισμό", available at: https://www.protagon.gr/ epikairotita/i-airbnb-anakoinwnei-stoixeia-kai-gia-ellada-kai-pairnei-thesi-gia -ton-ypertourismo-44341604912 (accessed on 3 October 2022).
Povich, E. (2014), "How Governments Are Trying to Tax the Sharing Economy. Governing", available at: http://www.governing.com/news/headlines/ governments-are-trying-to-tax-new-things.html (accessed on 3 October 2022).
Prassl, J., and M. Risak (2016), "Uber, Taskrabbit, & Co: Platforms as Employers? Rethinking the Legal Analysis of Crowdwork", *Comparative Labor Law & Policy Journal*, Oxford Legal Studies Research Paper No. 8/2016, available at SSRN: https://ssrn.com/abstract=2733003.
Rainnie, A., and B. Ellem (2006), "Fighting the Global at the Local: Community Unionism in Australia", Global Futures for Unions Conference, Cornell University Press, New York.
Ranchordas, S. (2015), "Does Sharing Mean Caring? Regulating Innovation in the Sharing Economy", *Minnesota Journal of Law, Science & Technology*, **16**(1), 413–475.
Rauch, D., and D. Schleicher (2015), "Like Uber, But For Local Government Policy: The Future of Local Regulation of the 'Shared Economy'", Working Paper No 21, Marron Institute of Urban Management, New York University, New York.
Rinne, A. (2015), "The Year of the Sharing Economy for Cities?", World Economic Forum, available at: https://www.weforum.org/agenda/2015/01/ 2015-the-year-of-the-sharing-economy-for-cities/ (accessed on 3 October 2022).
Rinne, A. (2018), "The Dark Side of the Sharing Economy", World Economic Forum, available at: https://www.weforum.org/agenda/2018/01/the-dark-side -of-the-sharing-economy/ (accessed on 3 October 2022).
Roberts, Y. (2018), "The Tiny Union Beating the Gig Economy Giants", *The Guardian*, available at: https://www.theguardian.com/politics/2018/jul/01/ union-beating-gig-economy-giants-iwgb-zero-hours-workers (accessed on 3 October 2022).
Rubery, J., A. Keizer and D. Grimshaw (2016), "Flexibility Bites Back: The Multiple and Hidden Costs of Flexible Employment Policies", *Human Resource Management Journal*, **26**(3), 235–251.
Santolli, B. J. (2017), "Winning the Battle, Losing the War: European Cities Fight Airbnb", *The George Washington International Law Review*, **49**(3), 673–709.
Schor, J. B. (2011), *True Wealth: How and Why Millions of Americans are Creating a Time-Rich, Ecologically Light, Small-Scale, High-Satisfaction Economy*, New York, NY: The Penguin Press.
Schor, J. (2014), "Debating the Sharing Economy", Great Transition Initiative, available at: http://www.greattransition.org/publication/debating-the-sharing -economy (accessed on 3 October 2022).
SEV (2017), "Οικονομία και Επιχειρήσεις, Special Report – Το Μέλλον της Εργασίας", Τεύχος 13, 13 Οκτωβρίου, available at: www.sev.org.gr/Uploads/

Documents/50583/SPECIAL_REPORT_MELLON_ERGASIA.18_10_2017 _final.pdf (accessed on 3 October 2022).

Sheingate, A. (2003), "Political Entrepreneurship, Institutional Change, and American Political Development", *Studies in American Political Development*, **17**(2), 185–203.

Sotiropoulos, D. A. (2017), "Reform Dynamics in Greek Democracy Today: Stagnation and Reform in Rule of Law, Mass Media and Social Inclusion", Friedrich Ebert Stiftung, July, available at: https://library.fes.de/pdf-files/bueros/athen/13572.pdf (accessed on 3 October 2022).

Sotiropoulos, D. A. (2012), "The Paradox of Non-Reform in a Reform-Ripe Environment: Lessons from Post Authoritarian Greece", in S. Kalyvas, G. Pagoulatos and H. Tsoukas (eds), *From Stagnation to Forced Adjustment: Reforms in Greece 1974–2010*, London: Hurst, pp. 9–30.

Thompson, P. (2003), "Disconnected Capitalism: Or Why Employers Can't Keep Their Side of the Bargain", *Work, Employment and Society*, **17**(2), 359–378.

Vandaele, K. (2018), "Will Trade Unions Survive in the Platform Economy? Emerging Patterns of Platform Workers' Collective Voice and Representation in Europe", Working Paper 2018.05, European Trade Union Institute, Brussels.

Visvizi, A. (2016), "Greece, the Greeks, and the Crisis: Reaching Beyond 'That's how It Goes'", available at: https://www.carnegiecouncil.org/ (accessed on 3 October 2022).

Voskeritsian, H., P. Kapotas, A. Kornelakis and M. Veliziotis (2019), "The Dark Side of the Labour Market: Institutional Change, Economic Crisis and Undeclared Work in Greece during the Crisis", in Pulignano, V. and F. Hendrickx (eds), *Employment Relations in the 21st Century: Challenges for Theory and Research in a Changing World of Work*, Kluwer Law International B.V., Chapter 15, n.p.

Zambarloukou, S. (2006), "Collective Bargaining and Social Pacts: Greece in Comparative Perspective", *European Journal of Industrial Relations*, **12**(2), 211–229.

4. Regulating the gig economy: promises and limits of social dialogue in Switzerland

Jean-Michel Bonvin, Nicola Cianferoni and Luca Perrig

INTRODUCTION

In Switzerland, labour market regulation takes place mostly through collective bargaining and social dialogue, based on decentralized and consensual relations between trade unions and employers' associations that are traditionally organized by economic sectors. Labour law provisions are minimalist, ensuring that actual working conditions are mainly regulated by actors closest to the ground (Bonvin 2007, Bonvin and Cianferoni 2013). Thus, in the Swiss context, workers' capacity to organize and to impact collective bargaining processes makes a real difference. In sectors where trade unions are well organized at branch level, they are also able to collectively negotiate better working conditions, most notably in terms of wages and working time, while in unorganized sectors working conditions are negotiated at the micro level between individual workers and employers, the only limitation being to comply with existing minimalist legal provisions. For a long time, this industrial relations configuration has proved beneficial for all stakeholders, as collective bargaining was characterized by a desire to find consensual or so-called "win–win" solutions, which could also guarantee social peace and stability: in line with the Fordist compromise (Supiot 2012), subordination at work was exchanged against enhanced material security, contributing to a virtuous circle of economic prosperity and improved social justice.

This model is being challenged by financial and economic globalization and the ensuing tendency to deregulate labour markets. Win–win strategies are more difficult to implement under such circumstances. In many cases the workers' position has been weakened whereas the employers' bargaining power has increased (Widmer 2007), leading trade unions to look for new structures and modalities of organizations (Oesch 2011). The emergence of the gig economy as a largely unorganized sector, where employers are not pushed to collectively bargain working conditions, comes as a further challenge to the Swiss model of social dialogue; such development is sometimes interpreted as a factor potentially disrupting the existing model of industrial relations. Working in the gig economy indeed implies significant differences compared with traditional forms of economic activity. The most obvious relates to the employment status, as gig workers are often hired as self-employed in sectors that traditionally employ their workers with an employee status. This affects gig workers' access to a full spectrum of labour rights and social protection (Pärli 2019). Besides, the gig economy's reliance on automated management (Lee et al. 2015) also raises questions about the ability of established social dialogue and collective bargaining patterns to adequately tackle such issues. Against these trends, trade unions are experimenting with new strategies: seeking to shift labour disputes from the company level to the political arena (Widmer 2007, Oesch 2011); inventing new ways to organize in order to involve gig workers in their conventional organizational structures and activities or, alternatively, to create new *ad hoc* organizations for them; or resorting to court litigation as a means to contest working conditions within the gig economy. In short, the gig economy has developed as a field of political and ideological struggle, where a multitude of actors are involved.

This chapter investigates how the gig economy impacts the Swiss industrial relations model based on class consensus and the principle of subsidiarity and what strategies Swiss trade unions implement in this context. To answer these questions, it takes a twofold perspective: first, a focus on the issue of digitalization in Switzerland and the overall debates it has produced among social partners and policy-makers; second, a snapshot on three case studies in sectors where the gig economy seems to be particularly flourishing in Switzerland, namely the taxi industry, bike deliveries and cleaning services. The first part of the chapter investigates the legal and political debates surrounding the emergence of the gig economy in Switzerland. Data were gathered from academic literature, official documents (notably reports produced by Swiss Government agencies, often in reaction to MPs' postulates at the Swiss parliament), national and local media articles and public media, as well as in-person interviews with key actors, namely five policy-makers, five academics and five high-rank social partner representatives. The second part relies mostly on interviews with key actors of the investigated sectors, more specifically seven drivers, three social partners (two trade unionists and one representative of a traditional taxi drivers' association) and two platform managers' representatives in the ride-hailing sector; six bike couriers, three platform managers and three trade unionists in the delivery sector; a manager and four cleaners in the cleaning sector.

1. DIGITALIZATION AND THE GIG ECONOMY IN THE SWISS PUBLIC DEBATE

Few people seem to engage in platform-mediated gig work in Switzerland, at least those who fully depend on gig work for their living are very limited in number (Federal Council 2017a). Statistical data were for a long time scarce and very inconclusive, as definitions of what is the gig economy diverge and most of it goes undeclared. State reports underlined the small size of gig work and its limited contribution to overall GDP, emphasizing the high percentage of dependent work (85 per cent) with no sign of decline (Federal Council 2017a). The first representative survey conducted by the Swiss Federal Statistical Office confirmed these statements: in 2019, only 0.4 per cent of the population said that they had carried out work via internet-mediated platforms in the past 12 months.[1] In this respect, the situation in Switzerland seems very close to that in other European countries (Ellmer et al. 2019), although the impact of the Covid-19 pandemic still needs to be thoroughly investigated.

Even though the precise size of the gig economy cannot be inferred from available research, it is possible to identify those sectors in which gig work is most prominent in Switzerland, namely transport of persons and delivery. A fierce competition has been taking place in both sectors since the emergence of digital platforms. Wages among gig workers seem to be considerably lower than local standards, and they often lack access to minimal social protection (Pärli 2016a). In the transportation sector, Uber's arrival in Switzerland in 2014 boosted competition with implications for traditional taxi drivers too. Deliveries via online platforms are also gaining importance in Switzerland. Multinational companies such as Deliveroo, Foodora, Amazon and UberEats have been remarkably absent until November 2018, when UberEats announced its launch of activities in Geneva. Meanwhile, multiple local platforms had emerged and the competition among them had become fierce. They mainly offer food deliveries, but many are trying to expand their services to groceries and parcel deliveries. Although their importance is not precisely quantified, they are abundantly discussed in the media due to their high visibility in the public space. The extent of platform cleaning services is also difficult to evaluate. A lot of it is undeclared and takes place behind closed doors.

Despite the uncertainties about its size, and partly due to the high visibility of gig workers during the pandemic period (especially bike deliveries in times of lockdown), the working conditions within the gig economy have attracted much public and media attention. The platform economy is indeed perceived by some, mostly trade unionists and members of left-wing parties, as a threat to the standard employment relationship (SER) that was set up after World War II and the guarantees it offered to workers in terms of employment security, material well-being and access to social protection (see Stanford 2017). This is mainly related to the self-employed status granted to most gig workers until recent years, which questions the very foundation of labour law and social protection as they are anchored in the wage earner status. Opponents to the gig economy thus emphasize the precariousness that is often associated with so-called gig jobs. Working for digital platforms means that the regularity of wages and the duration of working time are not guaranteed. It is thus emphasized how gig workers can be highly exposed to demand fluctuations, although this exposure varies along the worker's qualification and the geographical place (Portmann and Nedi 2015). Their undermined employment status also adds to the difficulties faced by labour inspectors when trying to assess the implementation of labour law and social policies in the gig economy. All in all, it involves a transfer of social risks from the employer to the individual worker, with significant consequences for the degree of protection against social risks and for the overall financing of the welfare state (Pärli 2019).

These critical views go in the direction suggested by Stanford (2017) of an erosion of the SER in the present context, to which the emergence of the gig economy further contributes. In the same critical line, some studies show that employment in the gig economy is motivated by the will to move out of unemployment rather than by the attractiveness of time flexibility, the interest in the gig activity or the possibilities to network or acquire new skills (Mattmann et al. 2017). By contrast, other studies insist on the opportunities offered by the gig economy and its flexibility, emphasizing its potential for creating economic activities and jobs. They claim that an excessive regulation could prevent the benefits of the gig economy and advocate for solutions that would preserve the flexibility offered by the gig economy (McKinsey Global Institute 2018). At the time of their emergence, platforms were employing workers as independent contractors with very limited access to social protection and protection from labour law. The public debate mainly focused on Uber and revolved around the key issue of the employee status, which in Switzerland, as elsewhere, also conditions to a large extent the access to social benefits (Riemer-Kafka and Studer 2017). On the one hand, many elements seem to go in the direction of a subordinate position of gig workers, which would justify the adoption of an employee status (Pärli 2016a – see also Meier and Pärli 2019 for milestones case law at EU level). For instance, the platform controls workers' performance via a rating system; the driver is dependent on the platform on the logistical and economic point of view; for most drivers Uber represents their sole ride provider. Other arguments point towards a self-employed status: three parties are involved in the contract (the customer, the platform and the worker), while only two parties are involved in a traditional labour contract; drivers are technically free to accept or decline the rides; they use a private car and are formally free to work whenever they want. These lines of argumentation are used by Uber to contest its employer status and be considered instead as a mere intermediary (Kahil-Wolff 2017). They are, however, contested by many interviewees, such as trade unionists and left-wing policy-makers, who remind that historically the independent contractor status has been designed for highly qualified workers – lawyers, doctors, dentists and so on – who prefer to choose freely for themselves their own social coverage. This is not the case of less qualified gig workers who often depend on one provider. Even if they are formally free to organize their work, the platform actually controls their labour process via algorithmic management (Lee et al. 2015). The similarities are strong with the "putting-out" system that existed before the industrial revolution (Stanford 2017). Thus, the status of gig workers is at the core of the debate among Swiss policy-makers, social partners and lawyers, which takes place along lines very similar to what can be observed in Germany or the UK (see chapters 2 and 5 in this volume).

The Swiss Government (that is, the "Federal Council") published various reports on the issue of digitalization and the gig economy, which can be considered as milestones in this debate (Federal Council 2017a, 2017b, 2021). The first two reports published in 2017 tackled the overall issue of digitalization, respectively the general economic conditions and the opportunities and risks for employment and working conditions. In both reports, the gig economy is not identified as a key issue in the Swiss digitalization debate, but as a topic among others that contributes to the current transformations of the world of work. Other issues such as automation, digital skills, flexible working conditions and unemployment are considered more important. Indeed, they appear more likely to create structural labour market changes since automation, for example, can replace the workforce and deeply change the existing jobs profiles. Still, the second report pays more attention to the status of gig workers. It discusses the distinction between dependent and independent workers from the point of view of law, practices and jurisprudence. The role of social partners is also mentioned, but only in broad terms and in connection with the general frame of digitalization – that is, without referring specifically to platform and gig work, touching upon issues that are already part of social partners' competencies: vocational training, collective labour agreements, and labour market control and surveillance. At this initial stage, there seems to be no need to consider gig work specifically and to develop *ad hoc* measures. Also, the Swiss Government tends to endorse an optimistic view about digitization and its potential, seeing it more as an economic opportunity than a risk of increasing workers' precariousness and stating that the Swiss model of social dialogue is an adequate tool to tackle all these issues.

A report commissioned by the State Secretariat for Economic Affairs (SECO) argues that the gig economy creates opportunities that social partners could seize for engaging in social dialogue (Meier, Pärli and Seiler 2018). In particular, the following issues seem to have gained significance for new forms of work organized through platforms: occupational health and safety, working time, flexibility, and workers' participation. Nevertheless, this does not result in a reinforcement of social dialogue within the gig economy. Platform managers are notoriously silent, except for some very rare interviews in newspapers where they seek to legitimize their practices against regular accusations of evasion from standard labour law, emphasizing the benefits of the gig economy for consumers and the inevitable reluctance met by disruptive business models at the time of their emergence. Most of them do not belong to any employer organization, because they consider themselves "intermediaries" and not "employers", while existing employers' associations advocate for maintaining the *status quo* and not adopting new regulations in the digital sector in the name of business freedom. By contrast, trade unions argue that the Swiss Government does not adequately address the issue of precariousness generated by the gig economy: some interviewees even consider that public authorities take the gig economy as an opportunity to promote further deregulation. They also accuse platforms to steal every year more than 60 million Swiss francs from gig workers through unpaid social contributions. These conflictual views seem difficult to reconcile and, at the time of writing (February 2022), prospects for adopting new labour regulations at legislative level are quite limited.

There are thus opposing views about the gig economy and its potential opportunities and risks for the whole society. On the one hand, it is claimed that platforms create new jobs for people who have difficulties in accessing the labour market and allow them to enjoy more temporal and organizational flexibility. On the other hand, those new jobs are perceived to increase the precariousness in the labour market and create a new working underclass. The ongoing debate over worker status is representative of this tension. In this context, trade unions focus their action on ensuring that gig workers are considered "employees". They do not strive for a reform of existing labour law provisions, but struggle against what they perceive as a misclassification of gig workers as "self-employed". To advance their views, they deploy strategies that seem to be at odds with the traditional Swiss model of industrial relations characterized as peaceful and consensual. An open conflict escalated for instance in the canton of Geneva, where in 2019 the local government decided to ban the services of UberX and UberEats. Such intervention of public authorities in labour market regulation is rare, but it has been advocated by trade unions as an alternative to collective bargaining, which could hardly take place in the gig economy. This was mainly due to difficulties to recruit members among gig workers and to the refusal of platforms to consider trade unions as legitimate stakeholders. This difficulty was further reinforced by the self-employed status that prevented workers' collective organization in the name of competition law (Meier and Pärli 2019). At the time of writing, collective labour agreements, the traditional tool of social dialogue in Switzerland, are barely used to regulate the gig economy, although few attempts can be observed in this direction. Trade unions are actively engaged in campaigns to denounce the gig economy and its working conditions, but with little impact as yet on grassroots actors.

The increasing awareness about the risks of gig work (boosted by numerous media articles and TV and radio broadcasts), together with the augmented pressure from some MPs and social partners, progressively led the Swiss Government to initiate a discussion about gig workers' employment status. The creation of a "third status", lying somewhere in between the self-employed and the wage earner, has been at the centre of a polarized debate between left- and right-wing parties. Finally, it has not been adopted by the Swiss Government (Federal Council 2021). The political discussion around this issue did not involve gig workers and platform managers themselves, and mostly consisted in public declarations and legal debates. Trade unions and employers' umbrella organizations commissioned legal scholars in order to bring evidence on the most appropriate employment status for gig workers, with a view to defending their respective positions (Pärli 2016a, 2016b; Kahil-Wolff 2017; Riemer-Kafka and Studer 2017). Studies commissioned by trade unions emphasized that gig workers have a subordinate relationship with their platform and should thus be considered as workers and enjoy labour law protection and access to social rights. By contrast, scholars commissioned by employers' associations or directly by platforms insisted on the urgent need to update Swiss labour law in order to boost the gig economy potential for creating new jobs and new markets. In this line, one interviewee advocated for the adoption of a third and hybrid status moving beyond the "outdated dichotomy wage earner/self-employed". In her view, such status would ensure minimal and adequate social protection, although inferior to that of wage earners, while taking into account gig workers' desire for more flexible employment relationships.

Parallel to this controversial public debate, there was also an increasing court activity about the status of gig workers and its implications. The main issues revolved around the employment relationship between a digital platform and a gig worker and its implications for social protection. In such court litigations, trade unions supported workers' claims, but they did not have the most important role. Indeed, government agencies in the field of social insurances played a decisive part in these legal cases. In the Swiss social protection system, the State pension fund (AVS) and the State accident insurance (SUVA) are entitled to decide whether a worker is self-employed or an employee, based on the economic activity and the type of employment relationship. They are able to decide, for instance, whether Uber should be considered an "employer" who has to fulfil its duties and contribute to workers' old-age pensions and protect them against professional injuries. When such decisions by the AVS and the SUVA are contested, decisions are to be made by civil courts on the basis of existing legislative provisions. What is at stake, then, is the proper implementation of labour law, not whether it should be reformed.

The legal debate focused on Uber workers' status; it went on for years and in the last two to three years many cantonal jurisdictions in the field of social insurances made decisions that Uber workers could not be considered self-employed (Pärli 2022). However, these case law decisions do not change the law itself, meaning that new decisions have to be made for each company so that the employee status can be recognized for their workers. In other words, every case needs to be examined by the courts in order to determine whether the concerned workers can be granted an employee status or not. Even though case law tends to recognize gig workers as wage earners (at least in the transportation and delivery sectors), this does not mean that platform companies are prepared to comply with their duties as employers. Rather, they are experimenting with new strategies to maintain low production costs and avoid the costs foreseen by labour law and social protections, such as subcontracting or other forms of externalization of labour costs (Pärli 2022).

Following the controversial public debate and the various court litigations in past years, the Swiss Government issued a report (Federal Council 2021) stating that no labour law adaptation was necessary and that the existing provisions allowed settling all disputes over gig workers' status. More specifically, it was advocated to implement a "flexi-test" to decide on a case-by-case basis, following existing legal criteria, whether a gig worker is an employee or self-employed and to what social rights he or she is entitled. At first sight, this solution seems quite close to what the EU proposal directive, issued in December 2021, advocates as a strategy against worker misclassification as self-employed (European Commission 2021). Contrary to the European Union, however, Switzerland does not envision implementing an employment presumption in all platforms exercising control over their employees; rather it contends that this issue is to be solved on a case-by-case basis, either via an agreement between social partners (in line with the Swiss industrial relations tradition) or, failing such an agreement, via a judicial process. This clearly demonstrates that no reforms or adjustments of the existing labour law are planned.

All in all, the legislative *status quo* was interpreted as the best compromise or, one could say, the lesser evil. On the one hand, further labour market regulation as advocated by left parties and trade unions was refused in the name of preserving business opportunities created by the gig economy; on the other hand, the creation of a hybrid or third status supported by employers' associations and right-wing parties was rejected as it was perceived as a first step towards the dismantlement of labour law. The compromise solution was that existing labour law was sufficient to ensure access to labour and social rights to gig workers who could demonstrate their employee status through the "flexi-test" or who would be recognized as employees in court decisions.

At political and judicial level, the debate focused on worker status. This implies that issues of job quality and working conditions have not been tackled at this level and need to be solved at sector and firm level, which is very much in line with the Swiss configuration of industrial relations. It is thus important to investigate what happens at this meso (sector) and micro (firm) level. This is the objective of the next section.

2.   ACCOMMODATING OR OPPOSING THE GIG ECONOMY? PRACTICES OF COLLECTIVE ORGANIZATION AT LOCAL LEVEL

As mentioned in the introduction of this chapter, social dialogue at sector and firm level plays a crucial role in the Swiss model of labour market regulation, where a significant part of the protection enjoyed by workers comes from collective labour agreements. However, such social dialogue is made difficult in the gig economy, both because the sector is largely unorganized due to the various forms of fragmentation at work (Heiland 2020; see also the introduction of the volume) and because gig workers have been (mis-)classified as self-employed workers (in Switzerland as in most other countries, legal provisions framing the right to social dialogue and collective bargaining are enshrined in labour law, which applies only to an employee–employer relationship). Contrary to a conventional employer, platforms thus have no obligation to provide workers' councils for example. Some commentators even suggested that gig workers' associations could be considered as forms of anti-competitive collusion that could be reprehensible under antitrust law (Cherry 2016), since they would distort market competition. Nevertheless, we have seen in the first part of this chapter that both the Swiss Government and civil courts consider that the issues of job quality and working conditions in the gig economy have to be solved at sector and firm level, where social partners should be involved according to the Swiss industrial relations configuration. This raises the question of how to create social dialogue from scratch in the absence of a legal framework supporting collective organization and representation.

In a first stage, most existing unions were hesitant about how to address gig economy issues, which can be understood with regard to their tradition. Longstanding trade unions have little experience with self-employed workers and have traditionally been reluctant to promote this status (Joyce and Stuart 2021). They are thus faced with a dilemma. On the one hand, if they build bridges towards self-employed workers they risk facing criticism for giving increased legitimacy to what many consider "grey zones" (Xhauflair, Huybrechts and Pichault 2018). That could indeed be interpreted as a kind of recognition of gig workers as self-employed. On the other hand, if they try to re-qualify these workers as employees, they must often do so without the workers' support. Indeed, many gig workers enjoy the flexibility provided by platforms; some prefer bearing the limitations linked to the self-employed status than seeing the platforms flee the city and lay off ("disconnect") their workforce as had already happened in many places. Such issues related to the employment status exacerbate trade unions' difficulties in recruiting gig workers. As an alternative strategy, unions may choose to turn to policy-makers to ask for a reinforced regulation of the gig economy that would strengthen their collective bargaining power, for example by asking the State to declare the implementation of a collective labour agreement compulsory not only for the contracting parties, but for all workers and companies of an economic sector (Bonvin and Cianferoni 2013).

At sector or firm level, social dialogue can take different shapes depending on the context in which the gig economy operates. One obvious example relates to the distinction between remote and on-location gig economy. Wood et al. (2018) showed, for example, that social dialogue in the remote gig economy emerges mainly through social networks. Regarding on-location platforms, Vandaele (2018) observes three main types of collective representation: a bottom-up approach (platform cooperatives or grass-roots unions), longstanding unions based on pre-existing structures, and freelancers' quasi-unions. In a similar line, Joyce and Stuart (2021) distinguish between longstanding mainstream unions, which tend to be politically moderate and privilege consensual social dialogue, and grass-roots unions endorsing more radical views and more conflictual action repertoires. In Switzerland, two complementary strategies are implemented: while UNIA, the main trade union in the country, strives to recruit and mobilize gig workers themselves (see the Uber and notime cases below), Syndicom (the trade union in the telecommunication and media sector) prefers social dialogue with conventional employers, trying to build a strategic alliance with them in order to counter the unfair competition by platforms, in the hope that this strategy will be supported by the State. We will see in our case studies, all concerned with location-based platform work, how the modalities and forms of collective action heavily depend on the existence (or not) of an already-established profession, and on the specific profile and motivations of gig workers.

The next subsections present three case studies in Switzerland. The first case study focuses on a platform for cleaning services. This platform is considered by most of its workers as an opportunity to exit the precarious working conditions prevailing in the cleaning sector. It thus illustrates a case where the business model of the gig economy is not challenged by workers, despite a recognized need for enhanced social protection. It showcases that the absence of support for setting up collective bargaining mechanisms can result in the absence of social dialogue in unorganized sectors. Our second case study focuses on one food-delivery platform in the canton of Bern. This case illustrates the struggle of gig workers in a sector where traditional workers are poorly organized. They thus have to mobilize on their own, with limited help from existing organizations and in the absence of political support. This comes, however, with the advantage that they are able to address issues that are specific to platform work. This results in a bottom-up approach to social dialogue, with subsequent difficulties to connect with existing organizations and to promote long-lasting regulation of the gig economy. Finally, the third subsection addresses the case of Uber in the canton of Geneva. It illustrates what may happen when platforms impinge on the market of a well-established profession. Taxi drivers, living on their wages, joined forces with a longstanding trade union to ask for a re-qualification of gig workers as employees, finally resulting in an intervention of the State and legislative reform. If existing trade unions succeeded in involving the State in the regulation of the gig economy, the challenge of connecting with grass-roots organizations and adequately tackling actual working conditions was not fully taken up.

2.1   The Case of Cleaning Services in Canton Vaud

For the last decade or so, the cleaning services sector has been invested by newly established platforms. SuperCleaner[2] has been operating since 2015. Its business model is different from other platforms in that it does not rely on a self-employed workforce. Indeed, all cleaners are hired as regular dependent workers: they pay taxes and social insurance contributions, and receive higher wages compared with traditional cleaning workers. However, the actual employer is not the platform, but the private household who uses the platform to hire the cleaner, usually for two or three hours per week. To implement this arrangement, a trust company has been established that represents legally the platform and each customer as an employer towards the cleaners. On their behalf the trust company pays salaries and social insurance contributions. Even though forty workers are directly employed by the platform (direction, IT developers, call centre workers for the customer care, lawyers, adminis-trative staff, etc.), they do not include about a thousand cleaners working for or through the platform. These are mostly female and two thirds of them are migrants. Some of them worked for traditional cleaning compa-nies before. When they decided to change their employer, the main moti-vation was that they would receive higher wages, enjoy greater autonomy at work and be able to choose their regular clients. The high degree of autonomy at work explains why the cleaners accept the lack of social protection. All in all, the cleaners interviewed seemed to consider that the platform was offering better working conditions than traditional cleaning companies and that it represented a significant improvement vis-à-vis undeclared work for private households that some of them were doing before joining SuperCleaner. One manager of the platform explains his plan to increase the salaries for the next ten years in order to keep the platform attractive, as well as his attempts to secure a complementary private pension scheme for part-time workers. These private pension schemes are a crucial component of retirement pensions in Switzerland and do not apply when the employer hires a worker for less than eight hours per week.

This does not mean that the cleaners were happy with the limited social protection provided by the platform. On the contrary, interviewees often complained about this, emphasizing for instance that they had to behave carefully to ensure they did not become ill and lose all their income. They also pointed to the pension issue, many of them stating that they would leave Switzerland after retirement to be able to make a decent living. However, this did not result in any form of collective action. In a poorly organized sector, cleaners felt that such collective claims would raise managers' discontent and they feared losing their jobs. In this case study, social dialogue did not play any role. There was no grass-roots mobilization and very limited involvement of longstanding trade unions. Workers themselves did not seriously contest their working conditions, although they admitted that improvements would be welcome. This case shows the opportunities and limitations of the gig economy when a sector is unorganized and its functioning is left to market actors: despite the obvious need for enhanced protection, the absence of any kind of support impinges on the possibilities to organize collective action and negotiate working conditions.

2.2   The Case of Bike Delivery in Canton Bern

Notime is active in the bike delivery sector. Since its creation in 2014, this platform offers delivery services to businesses, typically restaurants or parcel delivery companies. The company started in Zurich before expanding to eight main cities in Switzerland and employs hundreds of couriers. Its workforce is composed mostly of young male students. It manages its workforce via the usual devices of algorithmic management: customer ratings, online shift-picking, financial incentives, and automated dispatching. In a first stage, the company hired self-employed couriers on a piecework basis like most other platforms. Workers did not benefit from the social protection that comes with an employment status such as a pension fund, casualty or unemployment insurance. Moreover, they had to pay a value added tax (VAT) to the State instead of paying social contributions. At the same time, couriers had to follow strict rules related to time schedules, dress code, handling of goods, or behaviour towards clients. Several couriers complained about such conditions, but this did not lead to any collective action up to 2017. The mobilization only started when the managers decided that workers would benefit from the wage earner status. This was part of a deal related to the takeover of the company by the State-owned Swiss Post, which was conditional upon the fact that notime managers would "regularize" or "legalize" their business model and comply with the regulatory requirements of the Swiss mail market.

The mobilization began with informal meetings among bike couriers who were concerned about the increased flexibility of working schedules that the managers wanted to impose in connection with the new labour contracts. As dissatisfaction grew, bike couriers started to hold regular meetings with the UNIA trade union local branch. The creation of an instant messaging group, where supervisors were excluded, allowed gig workers to communicate freely and come together despite very individualized working conditions. These meetings resulted in the identification of their main claims: being employed as dependent workers without wage reduction, being collectively represented through the creation of a workers' council, and getting more transparent information regarding the algorithmic management of the labour process. By asking for a workers' council, they wanted to ensure that their collective voice could be heard within the platform (Cianferoni, Perrig and Bonvin 2022) and they would have the opportunity to express their claims and push the management to improve their working conditions. Specific emphasis was placed on the opacity of algorithmic management and the need for more transparency, especially with regard to how their performance was rated and how this impacted their access to shifts or their wage.

The trade union had an important role at this early stage. Nonetheless, its importance declined when notime's managers refused to engage in negotiations if trade union representatives were included in the workers' delegation. At first, bike couriers did not accept this; in an attempt to put increased pressure on the platform they decided, together with the trade union, to organize a public demonstration. However, the strong reaction of the managers, who accused the trade union of manipulating bike couriers for purposes disconnected from the dispute, pushed them to accept coming to the negotiating table without the trade union.

In the end an agreement was found between notime managers and gig workers, which stated under what conditions all bike couriers would be considered employees from 1 January 2018. Everybody would receive a fixed hourly minimum wage for every working shift, including extra compensations for the use of a private bike and mobile phone. Moreover, a more transparent ranking system was to be introduced while working conditions related to uniform, communication, occupational safety and app improvements were put on the agenda for further negotiations. The agreement also provided for the creation of a workers' council composed of elected bike couriers, which would be independent from the platform but would not include the trade union. Its mandate consisted in negotiating new labour contracts providing higher salaries, implementing a more transparent performance-based bonus system, and finding an acceptable solution for the payment of all backdated social contributions (that is, those that were not paid while couriers were considered self-employed).

Admittedly, the mobilization took place with the support of an established trade union, but from the beginning to the end it unfolded mostly as a collective action conducted by self-organized gig workers. The trade union support was important in giving the couriers more confidence about their claims and helping them build their strategy. However, it was not involved in the collective bargaining process and, even though couriers were grateful for the support received, this did not translate into longer-term collaboration after the end of the mobilization.

As a matter of fact, gig workers' actions inside notime very much look like the self-organized mobilizations that took place among couriers of online food-delivery platforms in other European countries (Tassinari and Maccarrone 2020). The case of notime is specific in that there was an initial attempt at combining grass-roots and trade union action, but with very limited outcomes in the end, which illustrates the challenges faced by trade unions when it comes to recruit and involve gig workers in conventional organizational structures.

2.3   The Case of Uber in Canton Geneva

Uber has been at the forefront of the debates about the gig economy in Europe, and Switzerland is no exception. Since its arrival in 2014, the company has raised numerous controversies. In this subsection, we focus on Canton Geneva, whose government has been among the first to take action in this regard. When Uber arrived in Geneva, the transportation sector was composed of taxi and limousine drivers subjected to a common law, the Taxi Law, establishing who had the right to engage in the business and under which conditions. Most taxi drivers were employed by taxi operators and only a few were independent, but all of them had to pass an exam in order to get a professional driver's license, with a success rate of only 50 per cent. This licence granted them the right to use taxi lanes to conduct their business, but also imposed duties such as installing a tachometer on their vehicle in order to keep track of their speed and activity and monitor their compliance with safety legislation. No official union existed at that time and the sector was not subjected to any collective labour agreement, but the competition within the profession was regulated by several provisions of the Taxi Law.

In spring 2014, Uber launched its service UberPop in Geneva, with prices about 25 percent cheaper than regular taxi rides. The first concern among the established taxi drivers revolved around unfair competition; and in September 2014, they asked the local government to take action against the platform under the claim that its drivers operated out of the Taxi Law. It is thus an issue of unfair competition that initiated a year-long conflict between traditional taxi operators and Uber lawyers to clarify the place of online ride-hailing services. Following a complaint from taxi operators, the local government stated that Uber drivers had to hold a professional driving licence just like traditional drivers. The platform itself was considered a regular taxi operator and would have to ask for an appropriate licence in order to continue its operations. Uber refused to ask for such a license and continued its operations, and the service gained hundreds of clients week after week. After facing growing discontent from taxi drivers, the government decided to draft new legislation that would allow ride-hailing platforms to operate alongside traditional drivers. In its drafting, it had the potential to ban platforms, make enrolment of drivers difficult, or put a cap on the number of drivers in the canton, but these options were not retained. The so-called Lex Uber was devoted primarily to include platforms into the existing law regulating the market for taxi operators. It sought to appease taxi drivers by requiring Uber drivers to pass an examination and preventing them from using taxi lanes.

Uber drivers themselves were remarkably absent from the debates surrounding the platform implementation in Geneva. One simple reason is that there was no collective organization at the time. It is only after the new law passed that Uber drivers envisioned organizing. They started using numerous instant messaging group apps. On these chats, they discussed mainly day-to-day issues such as traffic jams or radars, but they also sought help in understanding the app. This resulted in the emergence of an Uber drivers' association, with the main objective to pressure Uber into implementing minimum fares for each ride that could compensate drivers for their trip on their way to the client. Another issue was related to customer ratings and their impact on drivers' eligibility to platform services. Drivers complained that their services were not consistently rated due to the diversity of practices from customers. All such issues, related to the day-to-day work of Uber drivers, could have been taken into account in the drafting of the new legislation, had the drivers been more vocal at the time. However, they were not taken up by the legislators whose main interlocutors in the Uber case were traditional taxi drivers.

This case study provides an example of a fragmented and disconnected collective action at local level. On the one hand, traditional taxi drivers defend their interests by calling for a ban on Uber activities; on the other hand, gig workers organize mainly to improve their working conditions but without contesting their working status. After Uber's arrival in Geneva, the government took measures to create a legal space for its business to flourish, but it was not able to anticipate issues arising from platform management, such as algorithmic accountability and personal data protection (Van Doorn and Badger 2020). To some extent, this case ran on two parallel tracks: on the one hand, the discussion between taxi operators and legislative actors; on the other hand an emerging, but quite hesitant, collective action at grass-roots level. Both are largely disconnected at this initial stage. This was also reflected in the twofold and to a large extent fragmented action by trade unions, which both tried to pressure the government into adopting legislation in favour of gig workers (focusing on worker status) and strove to recruit gig workers and involve them in collective mobilizations.

In the years following our empirical fieldwork, longstanding trade unions such as UNIA were able to better represent gig workers' interests and enforce the employee status for them, while preserving working conditions for traditional taxi drivers. They thus were successful, to some extent, in creating a bridge between the two forms of collective action taking place in Canton Geneva. This was achieved mainly through public campaigning in order to push local government to take regulatory action in the field of the gig economy. If legal issues (worker status) could be successfully tackled in this way, solutions have not yet been found for the more practical issues raised by gig workers, such as customer ratings or algorithmic management. Thus the gap between longstanding unions and new grass-roots movements has only been partially bridged in this case, as was confirmed by our interviewees.

3.   CONCLUSION

The case of Switzerland is characterized by a high stickiness of the existing model of industrial relations, based on social dialogue and an emphasis on the principle of subsidiarity. There is an overall consensus that the legal *status quo* is an acceptable compromise: therefore it is assumed that a labour law extension is not needed and that the existing provisions are appropriate to face the challenges represented by the gig economy. More specifically, the political debate on the gig economy mainly focused on the status of workers and its conclusion was that neither a reinforced regulation (in the sense of enhanced protection and rights for gig workers) nor the adoption of a third hybrid status (with reduced access to social rights vis-à-vis wage earners) were required; rather, it was asserted that the issue of gig workers' status could be properly addressed within the existing legal framework, if necessary with the support of judicial courts. All other issues, related to the day-to-day actual working conditions of gig workers, have received very limited attention in the political and legislative arena and are therefore to be addressed at sector and firm level.

This raises the question as to whether the conditions for an effective social dialogue between social partners are met within the gig economy. In the past few years, new forms of collective organization and social dialogue started to emerge despite adverse conditions – especially individualized working conditions, high risks of spatial and organizational fragmentation, and the pervasiveness of the self-employed status, all factors impinging on the implementation of social dialogue and collective bargaining mechanisms. Nonetheless, the case studies highlighted in this chapter illustrate that social dialogue can develop at meso (sector) and micro (firm) level, though in varying forms and with more or less impact and success along the different sectors and platforms investigated. Mobilizations were most visible in the bike delivery sector, where couriers appeared as pioneers in terms of collectively organizing and contesting working conditions.

Our empirical work allowed identifying multiple configurations of social dialogue. In some cases, longstanding trade unions were at the forefront, while in other cases the movement was more grass-roots; still other sectors saw no mobilization at all. These distinct forms of social dialogue in turn led to gig workers being more or less included in the social negotiation, the cleaning sector being even a case of absent social dialogue where gig workers had to implement their own individual strategies. It is worth noting that a joint long-term and successful involvement of gig workers and established trade unions in the negotiation process could be observed in none of the three case studies. How, then, can we account for such diversity of social dialogue configurations within the gig economy and for the difficulties to combine trade union action and grass-roots initiatives? Our case studies suggest three main reasons.

First, it appears that long-term collaboration between grass-roots movements and longstanding trade unions may be complicated by gig workers' economic dependency on the platform. Most platforms are aware of this and threaten to stop their operations if forced to employ directly their on-demand workforce. However, a strong distinction has to be made between workers in the ride-hailing sector and those doing food deliveries. Uber drivers typically rely heavily on their income from the platform and invest large amounts of money into the cars they use, whereas notime couriers engage in gig work for a shorter duration and invest almost no money in the process. It may well be the case, then, that Uber drivers are more reluctant to mobilize along established trade unions since the implications of being laid off are higher for them. Empirical studies, however, show that this factor does not prevent them from taking grass-roots initiatives and organizing collective mobilization.

Second and closely related is the distinction between local and multinational platforms. The threat of withdrawal is mainly used by multinational platforms that can more easily diversify their sources of revenues and raise large investments. By contrast, in our case study in the delivery sector, which focused on a mobilization taking place in a local platform, the threat of leaving the country was never formulated. This suggests that local platforms may be more likely to engage in social dialogue under the pressure of grass-roots movements, trade unions or organized professions.

Third, the existence of an organized profession or a trade-unionized workforce before the arrival of platforms allows longstanding trade unions or professional associations to have a firmer hold on the upcoming social dialogue. This was very clear in the transportation sector, where taxi drivers could unite with the help of trade unions in order to fight against the "uberization" of their activity. The opposite situation can be observed in the cleaning sector, where the limited degree of pre-existing collective organization is probably one of the main factors accounting for the lack of social dialogue. In such a context, the arrival of an alternative business model has not been met with strong resistance from cleaners in the conventional sector.

All in all, the Swiss model of industrial relations has shown high resilience in the face of potentially disruptive phenomena such as the emergence of the gig economy. The continuing relevance of the model is supported by most stakeholders in the political arena, who consider the legal *status quo* as a workable solution to address the challenges of the gig economy. However, the success of this model strongly depends on the ability of social partners at sector and firm level to implement an equitable social dialogue in which win–win solutions can be designed. Our case studies at meso (sector) and micro (firm) level suggest that the successful implementation of this model faces some difficulties in the specific context of the gig economy. Developing workers' capacity for collective organization and mobilization appears as a prerequisite for the enhancement of social protection and social dialogue within the gig economy. This certainly requires a stronger support from public authorities, for instance granting additional social and participation rights, as well as innovative and more inclusive strategies from longstanding trade unions with a view to establishing more balanced power relationships between employers and workers of the gig economy.

NOTES

1. "Internet-mediated platform work is not very common in Switzerland", press release of the Swiss Federal Statistical Office, 19 May 2020.
2. The name of the platform is anonymized.

REFERENCES

Bonvin, J.-M. (2007), "Corporate Social Responsibility in a Context of Permanent Restructuring: A Case Study from the Swiss Metalworking Sector", *Corporate Governance: An International Review*, **15** (1), 36–45.

Bonvin, J.-M., and N. Cianferoni (2013), "La Fabrique du Compromis sur le Marché du Travail Suisse. Évolutions et Défis Actuels", *Négociations*, **20** (2), 59–71.

Cherry, M. A. (2016), "Beyond Misclassification: The Digital Transformation of Work", *Comparative Labour Law and Policy Journal*, **37** (3), 544–577.

Cianferoni, N., L. Perrig and J.-M. Bonvin (2022), "When Voices from Below are Heard: The Case of a Swiss Online Food-delivery Platform", in A. Wilkinson, T. Dundon and S. Brooks (eds), *Missing Voice? Worker Voice and Social Dialogue in the Platform Economy*, Northampton, MA, USA and Cheltenham, UK: Edward Elgar Publishing, pp. 195–215.

Ellmer, M., B. Herr, D. Klaus and T. Gegenhuber (2019), "Platform Workers Centre Stage! Taking Stock of Current Debates and Approaches for Improving the Conditions of Platform Work in Europe", Working Paper 140, Hans-Böckler Stiftung, Düsseldorf.

European Commission (2021), "Proposal for a Directive of the European Parliament and of the Council on Improving Working Conditions in Platforms", COM(2021) 762 final.

Federal Council (2017a), *Conséquences de la Numérisation sur l'Emploi et les Conditions de Travail: Opportunités et Risques*, Berne.

Federal Council (2017b), *Rapport sur les Principales Conditions-cadre pour l'Economie Numérique*, Berne.

Federal Council (2021), *Numérisation – Examen d'une Flexibilisation dans le Droit des Assurances Sociales ("Flexi-Test")*, Berne.

Harris, S. D., and A. B. Krueger (2015), "A Proposal for Modernizing Labor Laws for Twenty-First-Century Work: The 'Independent Worker'", The Hamilton Project, Discussion Paper, December.

Heiland, H. (2020), *Workers' Voice in Platform Labour*, Study No. 21, Hans-Böckler-Stiftung, July.

Kahil-Wolff, B. (2017), *Der AHV-rechtliche Beitragsstatus von in der Schweiz tätigen Uber-Fahrern Gutachten zu Händen von Uber Switzerland GmbH*, Université de Lausanne.

Joyce, S., and M. Stuart (2021), "Trade Union Responses to Platform Work: An Evolving Tension between Mainstream and Grassroots Approaches", in J. Drahokoupil and K. Vandaele (eds), *A Modern Guide to Labour and the*

Platform Economy, Northampton, MA, USA and Cheltenham, UK: Edward Elgar Publishing, pp. 177–192.

Lee, M. K., D. Kusbit, E. Metsky and L. Dabbish (2015), "Working with Machines: The Impact of Algorithmic and Data-driven Management on Human Workers", Proceedings of the 33rd Annual ACM Conference on Human Factors in Computing Systems, Heinz College Carnegie Mellon University, pp. 1603–1612.

Mattmann, M., U. Walther, J. Frank and M. Marti (2017), "L'évolution des Emplois Atypiques Précaires en Suisse" [Evolution of Atypical Precarious Jobs in Switzerland], Bern, Ecoplan, retrieved from: www.newsd.admin.ch/newsd/message/attachments/50256.pdf (accessed on 26 September 2022).

McKinsey Global Institute (2018), *The Future of Work: Switzerland's Digital Opportunity*, McKinsey & Company, retrieved from: https://www.mckinsey.com/ch/~/media/mckinsey/featured-insights/europe/the-future-of-work-switzerlands-digital-opportunity.ashx (accessed on 26 September 2022).

Meier, A., and K. Pärli (2019), "Commentary on Court of Justice of the European Union Judgments C-434/15 of 20 December 2017 (Asociación Profesional Elite Taxi v Uber Systems Spain SL) and C-320/16 of 10 April 2018 (Uber France SAS)", *Revue de droit comparé du travail et de la sécurité sociale*, **2**, 98–108.

Meier, A., K. Pärli and Z. Seiler (2018), "Le Futur du Dialogue Social et du Tripartisme dans le Contexte de la Digitalisation de l'économie", Etude établie sur mandat de la Commission nationale tripartite pour les affaires de l'OIT, Berne et Bâle, Geneva.

Oesch, D. (2011), "Swiss Trade unions and Industrial Relations after 1990: A History of Decline and Renewal", in C. Trampusch and A. Mach (eds), *Switzerland in Europe: Continuity and Change in the Swiss Political Economy*, London: Routledge, pp. 82–102.

Pärli, K. (2016a), *Gutachten "Arbeits- und sozialversicherungsrechtliche Fragen bei Uber Taxifahrer/innen"*, legal report commissioned by Unia, Bern and Basel, retrieved from: https://www.unia.ch/fileadmin/_migrated/news_uploads/2016-08-29-Gutachten-Arbeitsrecht-Sozialversicherungsrecht-Uber-Taxifahrer-innen-Professor-Kurt-P%C3%A4rli_01.pdf (accessed on 26 September 2022).

Pärli, K. (2016b), "Neue Formen der Arbeitsorganisation: Internet-Plattformen als Arbeitgeber", *Arbeitsrecht*, **4**, 243–254.

Pärli, K. (2019), *Arbeits- und sozialversicherungsrechtliche Fragen der Sharing Economy: Problemstellung und Lösungsansätze bei der Plattform-Erwerbstätigkeit*, Zürich: Schulthess.

Pärli, K. (2022), "Über Uber-Urteile und immer neue Uber-Geschichten...", *Schweizerische Zeitschrift für Sozialversicherung und berufliche Vorsorge*, retrieved from: https://szs.recht.ch/de/artikel/01szs1022abh/uber-uber-urteile-und-immer-neue-uber-geschichten (accessed on 21 February 2022).

Portmann, W., and R. Nedi (2015), "Neue Arbeitsformen – Crowdwork, Portage Salarial und Employee Sharing", in P. Breitschmid, I. Jent-Sorensen, H. Schmid and M. Sogo (eds), *Tatsachen – Verfahren – Vollstreckung*, Zürich: Schulthess.

Riemer-Kafka, G., and Studer, V. (2017), "Digitalisierung und Sozialversicherung – einige Gedanken zum Umgang mit neuen Technologien in der Arbeitswelt", *Revue Suisse des Assurances Sociales et de la Prévoyance Professionnelle*, **4** (17), 354–384.

Stanford, J. (2017), "Historical and Theoretical Perspectives on the Resurgence of Gig Work", *Economic and Labour Relations Review*, **3** (28), 328–401.

Supiot, A. (2012). *The Spirit of Philadelphia: Social Justice vs. the Total Market*, London: Verso.

Tassinari, A. and V. Maccarrone (2020), "Riders on the Storm: Workplace Solidarity among Gig Economy Couriers in Italy and the UK", *Work, Employment and Society*, **34** (1), 35–54.

Van Doorn, N. and A. Badger (2020), "Platform Capitalism's Hidden Abode: Producing Data Assets in the Gig Economy", *Antipode*, **52**, 1475–1495.

Vandaele, K. (2018), "Will Trade Unions Survive in the Platform Economy? Emerging Patterns of Platform Workers' Collective Voice and Representation in Europe", Working Paper, Brussels: ETUI.

Widmer, F. (2007), "Stratégies Syndicales et Renouvellement des élites: le Syndicat FTMH Face à la Crise des Années 1990", *Swiss Political Science Review*, **13**, 395–431.

Wood, A. J., V. Lehdonvirta and M. Graham (2018), "Workers of the Internet Unite? Online Freelancer Organisation among Remote Gig Economy Workers in Six Asian and African Countries", *New Technology, Work and Employment*, **33** (2), 95–112.

Xhauflair, V., Huybrechts, B. and F. Pichault (2018), "How Can New Players Establish Themselves in Highly Institutionalized Labour Markets? A Belgian Case Study in the Area of Project-Based Work", *British Journal of Industrial Relations*, **56** (2), 370–394.

5. Weakening worker protections? Uncovering the gig economy and the future of work in the UK

Tom Montgomery and Simone Baglioni

1. INTRODUCTION

The connection between the development of online technology and its implications for labour markets has been the focus of attention from scholars for almost two decades (Autor 2001). This has brought significant disruption to particular sectors and has implications not only for workers but also for existing employers in the sector, for the trade union movement, and for policymakers who must grapple with the consequences of such changes for the social protections that have thus far underpinned the relationship between employers and employees.

What we can begin to identify in the debate thus far, is a fusion between concerns regarding the impact of flexible labour markets and the disruptive potential of technology. In other words, there is some overlap between the concerns articulated by Standing (2011) regarding the emergence of a 'precariat' class of workers who have few employment rights and scarce employment security and the emergence of a 'cybertariat' class (Huws 2014) whose experiences, though materially similar, are specifically bound up in the disruption caused by technology and the power asymmetries that can be exacerbated in the markets for digital labour (Graham et al. 2017). Consequently, the emergence of the gig economy and the consequences that accompany its rise have fostered a growing critique of these new forms of work and the implications for the future living standards of workers across different national contexts (Scholz 2012, 2017; Wood et al. 2019).

In this chapter we offer an analysis that is multidimensional, in that the perspectives that are adopted cut across macro (policymaking), meso (trade unions and labour organisations) and micro (self-employed workers). Our chapter takes the following structure. First, we comprehend the UK context in which our study takes place and the particular challenges that the labour market environment presents for worker protections, and we outline the methods used in conducting our study. Next, we turn to the first stage of our findings, drawing upon the reflections elicited from our interviews to identify some of the specific issues confronting workers in the UK gig economy. We then turn to understanding the role of those involved in organising workers in the UK gig economy and their challenges in defending worker protections. Subsequently we look at the impact of policymaking on protecting these workers in the UK, before offering our conclusions.

2.   RESEARCH CONTEXT AND METHODS

In this chapter we illuminate the implications of the gig economy in a specific context, namely the UK, which has been experiencing relatively higher levels of income inequality (McGuinness 2017) compared with other European countries. The context of the labour market in the UK provides a key example of one where the types of employment practices found in the gig economy have been experienced by an increasing number of workers in recent years, particularly following the global financial crisis. Evidence of this can be identified when observing official statistics in the UK that measure the extent of employment insecurity through non-standard forms of employment such as 'zero hours contracts' (a type of casual contract where there are no minimum guarantee of hours provided by the employer and the employee is effectively 'on call') as illustrated by Figure 5.1.

One report commissioned by the Department for Business, Energy and Industrial Strategy (BEIS) provides a snapshot of that segment of the workforce currently composing the gig economy in the UK. Findings from the 2018 BEIS study revealed that in the UK courier services were the most frequent form of gig economy work (42%), along with transport services (28%) and food delivery services (21%). What the gig economy has come to symbolise for workers in sectors such as these is the opportunity for flexibility, to earn extra money through short-term opportunities, and thus brings for them a benefit. For others, however, the gig economy in the UK has come to be associated with a decline in the quality of employment, including the dilution of longstanding employment rights, an increase in employment insecurity (Gregg and Gardiner 2015) and

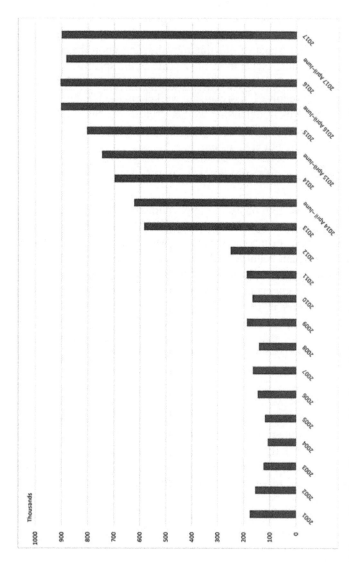

Figure 5.1   Number (thousands) of people in employment reporting they are on a zero hours contract

Source: Office for National Statistics (2018).

the stagnation of real wages (Resolution Foundation 2017). Employment insecurity and lower earning potential are therefore understood in the context of the proliferation of non-standard forms of employment such as zero hour contracts, but often missed in the discussion is the surge in self-employment in the UK in recent years (particularly those who are self-employed but with no employees of their own, see Figure 5.2), a phenomenon that needs to be appreciated if we are to truly understand the gig economy, as we shall set out later in this chapter.

Figure 5.2   *Self-employed with or without employees*

Source: Office for National Statistics (2018).

As part of our Swiss Network for International Studies (SNIS)-funded project into the 'Gig Economy and its Implications for Social Dialogue and Workers' Protection', we conducted a thematic analysis of interview data drawn from a range of actors (n=29) including policymakers (for example, parliamentarians from across central and devolved government), trade unions (for example, national level officials, regional level organisers) and labour organisations (for example, social economy organisations engaged in supporting gig economy and precarious workers) as well as gig economy workers (that is, self-employed delivery drivers based in a large urbanised area whose work depended on a single distribution company that coordinates the delivery of items to primarily household consumers from a range of well-known online shopping platforms) across the UK. Although we captured a variety of perspectives, we also made efforts to recruit employers and employer representatives to our study, but despite numerous attempts, these organisations were unwilling to participate, meaning that future research is necessary to capture the input of these actors. Most of our interviews were conducted either face to face or over the phone (Bryman 2016) with two participants electing to respond to our questions via email (Burns 2010). We focused on issues of social partners and workers' understanding of the impact of the gig economy, of workers' rights and working conditions, but we also explored how the gig economy offered challenges and opportunities for social partners. Our interview data was triangulated by observation of a gig economy strike, specifically the gig economy strike that took place in the UK on 4 October 2018 and which a member of our team attended, taking field notes.

The findings of our study are therefore built upon a foundation of different sources and enabled a rich insight into the issues generated by the gig economy for the prospects of social dialogue in the UK, a context where social partnership needs to be understood as embedded in a liberal market economy where bargaining takes place primarily at the level of the firm (Hall and Soskice 2001). Moreover, we also undertook an analysis of the wider UK policy context by examining a range of literature that included policy documents from the UK Government, Scottish Government, the Welsh Government and Northern Ireland. These documents were not sourced primarily because of any explicit reference to the gig economy (which is still nascent) but instead because they encompassed related issues such as employment security, pay and fair work, which, as we shall discover later in this chapter, connect to the development of the gig economy. We also examined those submissions made by employers, trade unions and other stakeholders to the Taylor Review of Modern Working Practices (see also Bales et al. 2018).

3. THE GIG ECONOMY AND THE IMPLICATIONS FOR WORKERS' PROTECTIONS

Almost all of our interviewees expressed concerns about the potential that existed for the acceleration of exploitation of workers, particularly in certain sectors and sub-sectors. In this section of the chapter, we summarise such concerns by focusing on two key aspects of the gig economy that emerge from our analysis, consistent with extant literature (Crouch 2019): that the gig economy has exacerbated the polarisation of the labour market (Goos and Manning 2007; Plunkett and Pessoa 2013; Goos et al. 2014), and that it has widened the power imbalance that exists in each employee/employer relationship. We note also that such an imbalance in favour of the employer, which is often a de facto relationship rather than one that is legally binding, is reflected through a range of issues spanning from time management to health and safety, wages, and gender relationships.

Concerns that the gig economy was a vector of poor employment practices were elicited across our interviews. For example, a trade union organiser in the hospitality sector explained to us that some of the businesses that were involved in using zero hours contracts could potentially be decent places to work but that there was, in his experience, a clear disparity regarding the terms and conditions of those working in precarious employment contracts and those in more standard forms of employment. Hence, the gig economy did contribute, according to this interviewee, to further the dualisation of the labour market brought in by the post-Fordist, global, economy. For this trade unionist the evidence of this disparity was reflected by the clear difference in the negotiations he had with established employers with secure employment on offer and the contrasting conditions found among those offering precarious work in the hospitality sector. Furthermore, the experience of insecurity was highlighted by a number of interviewees as indicative of the balance of power having shifted to the employer.

Moreover, it was clear for some interviewees that the lack of enforcement of basic employment rights was affecting some workers more than others. This is illustrated by the views elicited by one officer for a labour organisation, who explained that women were increasingly disadvantaged by the insecurity and low pay that they were finding in the gig economy. For this interviewee, the gender pay gap risked being exacerbated by the lack of awareness of how the conditions found in the gig economy disproportionately affect women:

> When people talk about the ills of the gig economy around holiday pay and maternity pay, all of these things affect men and women but for women the effect is much more intensified. Maternity pay is obviously most important for women. And 94% of lone parents are women so if you're working in the gig economy as a lone parent and you need childcare, flexible and affordable childcare isn't open to you. So how do you manage to care for your child? Women are more likely to be in poverty than men, more likely to be in in-work poverty than men, so women working in the gig economy further exposes them to all of this. (Labour Organisation: Parliamentary Officer)

For other interviewees, a key concern, still related to the widening power gap between employer and employee, was the sense that workers were losing the voice necessary to enforce their employment rights across different sectors. This was articulated for example by one national level organiser in the trade union movement who consistently emphasised the power asymmetries that were growing particularly among those occupations that had grown in the public consciousness as being part of the 'gig economy'. The organiser was adamant that technological change was being used as a cover for rolling back the relationship between employers and employees to one that was more suited to the nineteenth century than the twenty-first century. Emphasising the atomisation experienced by gig workers such as Uber drivers, the interviewee explained that he was increasingly concerned that the conditions being experienced by such workers were in danger of spreading to other parts of the labour market:

> I think one of the main effects that the gig economy has had is on wages and job security in the non-gig economy and is part of this wider loosening of the relationship between workers and the employer. (Trade Union: National Organiser)

The concern that there was a growing loss of connection between the employee and the employer was also reflected in some of our other interviews. For example, one of our participants from a labour organisation in the third sector explained that the workers he and his colleagues were supporting were navigating a space that was marked not only by diminishing rights but also by a loss of control over their work. As a consequence, he described his perception of the gig economy as almost entirely negative given that he was viewing it through the prism of the daily experiences encountered by workers engaging with his organisation.

Although he viewed these processes as cross-sectoral, he had witnessed particular problems emanating from the larger platform companies and the consequences for workers were that,

> [t]hey get paid less, they've got less secure work and their employment rights are not what they would have been as employees .... Uber is disrupting the industry and the knock[-]on effect is that their conditions and income [are] not what it could be or their control over their working environment such as working hours is not what it used to be. (Labour Organisation: Senior Executive)

Finally, another way to understand the gig economy as a widening gap in power relations among subjects is through the discussion of the self-employment dimension. One of the key selling points of the gig economy in social discursive confrontations is that the sector offers the opportunity to tap into the potential residing in individuals' entrepreneurial aspirations and skills. But we elicited from our interviews primarily concerns regarding the risks connected with the self-employment status. Part of these concerns can be traced back to the earlier section of this chapter, which illustrates how the growth in employment in the UK has been mirrored by a growth in self-employment. A number of interviewees expressed concerns that these new forms of self-employment were driven less by a new entrepreneurial spirit than by sheer necessity. To gain a firmer insight into some of these experiences, we interviewed self-employed drivers who were involved in the delivery arm of large online platform companies. One aspect that emerged during the interviews with the drivers was the sense that the companies with whom they were employed as 'self-employed' were consistently pushing down the earnings of the drivers, through frequent 'reviews', which was when drivers would be told by the company their new rate of pay. It was through such reviews that drivers began to realise the extent to which 'self-employment' in the gig economy had left them unprotected compared with other workers.

An example of the challenges faced by those workers participating in our study stemmed from an interview with a driver who had experienced a short period of illness and now was finding himself in dispute with the company to the point where he believed he would soon see his contract terminated. He explained the disciplinary system (Thompson 1967) in place for those who fell ill:

> Up until eighteen months ago if we phoned in sick and said 'I'm not well today', you would lose money, not make any money that day and you would get £150 fine for not being in. What they've introduced now, if you're off sick, it's a point system, so if you phone in sick you'll get three points and once you hit twenty[-]one points that's you suspended and that's when you have a meeting with the manager and it's totally up to them whether they get rid of you there and then … three points for being off sick each day. (Gig Economy Worker: Self-employed driver 1)

Issues of health and safety have become a mainstay of employment rights in the UK and beyond for decades, but our interviews indicated cause for concern that these standards were being eroded by the practices of companies hiring 'self-employed' drivers in the gig economy. This was highlighted in interviews, including with one driver who had found himself in dispute with managers because he had pointed out to colleagues that the number of hours they were driving in order to meet the delivery targets set by the company was potentially unsafe. On one occasion he confronted managers when a newly started driver was about to potentially work beyond the legal limit:

> I said to the manager, 'Are you telling him to break the law?' The general manager then came down and said, 'What are you doing getting involved, don't get involved, it's nothing to do with you, if he does it, he does it….' They came down the next day and told me that I would be getting punished for telling this to people. It's people's safety, regardless of whatever they say… and there are no signs up that you can only work this time, there is nothing like that. (Gig Economy Worker: Self-employed driver 2)

What was clear, even in such interactions as confrontational as the above, was that the drivers perceived all of the power in the relationship to reside with the company.

When disputes were raised, drivers were expected to speak to managers who acted almost in the role of intermediaries between the driver and the company and as such the employer–employee relationship was completely blurred both by the 'self-employed' status and the system that was set up by the company to receive grievances and complaints from their 'self-employed' suppliers. This was compounded by the fact that the trade union to whom some of the drivers belonged was not recognised by the company and consequently basic worker representation was absent.

This was a concern articulated by some of the trade union interviewees we spoke to, including one official involved in organising workers in the gig economy who was clear that what needed to be understood in order to progress the conditions of these workers was that, despite the technology and the 'self-employed' status of workers, there was actually an employer:

> That relationship is what we need to focus on rather than how do we make life better for self-employed gig workers. Every gig worker that you can unionise will have one boss. (Trade Union: Campaigns Officer)

4.   THE GIG ECONOMY AND ITS IMPLICATIONS FOR ORGANISING WORKERS

The responses of the trade union movement to the challenges posed by the gig economy need to be placed in the context of an ongoing struggle between the trade union movement and the Conservative-led governments at Westminster (where employment law is determined), who have held power for almost a decade. As well as being at the forefront of opposing austerity policies implemented following the financial crisis, the trade unions have found themselves in a contested relationship with central government that has culminated with the introduction of the Trade Union Act 2016;[1] this has increased regulations on trade union activism. These included, for example, a rule that, in order for industrial action to be considered valid, the ballot of members must reach a turnout of at least 50% of those who are entitled to vote (Ford and Novitz 2016). The introduction of such legislation by the Conservative government may suggest that there has been a spike in trade union radicalism, when in fact, historically speaking, the frequency of strike action taken by trade unions in the UK has fallen dramatically (see Figure 5.3). In fact, the figures on labour disputes in 2018 reflect the historical trend by revealing that the number of days lost to labour disputes in 2018 was the sixth-lowest annual total since records began in 1891 and there was a total of just 81 stoppages in 2018, the second lowest figure since 1930 (Office for National Statistics 2019).

In addition to this fall in trade union strike action, there has been (germane to the experience of other European contexts) a steady decline in trade union membership in the UK, albeit with a small increase in recent years (Department for Business, Energy and Industrial Strategy

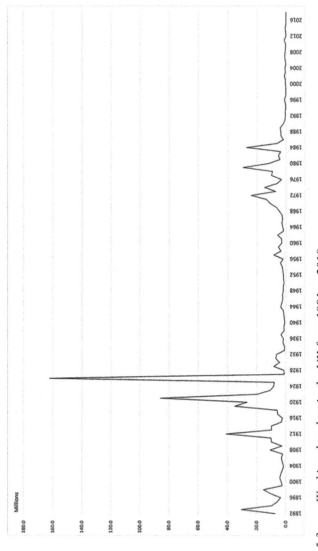

Figure 5.3   Working days lost in the UK from 1891 to 2018

Source: Office for National Statistics (2019).

2021). Moreover, and particularly crucial for understanding the challenge for trade unions, according to the same official statistics the trade union membership as it currently exists in the UK comprises typically older workers in professional roles who have permanent contracts and work full time (Department for Business, Energy and Industrial Strategy 2018). In other words, the composition of the trade union as it stands is not necessarily experiencing the types of practices that are emerging via the gig economy; and as such, if these practices spread and trade unions push forward efforts to recruit gig economy workers, they will need to be nimble enough to service a membership that has very different needs (see Hyman and Gumbrell-McCormick 2017).

One aspect of these campaigning efforts has been to pursue the recognition of Uber drivers as workers through the courts. This approach has yielded some significant success, most notably in the dispute between the GMB union and Uber in the UK[2] (Shead 2017; Aslam and Woodcock 2020), with UK-based Uber drivers now having formal representation following Uber's recognition of GMB.[3] Indeed, in the course of undertaking our interviews, reference was made by a number of our participants to the pursuit of legal action as one potentially fruitful route of enforcing the rights of workers. This included the view of one trade union organiser in the north of England who was involved in addressing the issues faced by those in the gig economy and pointed to the legal action that his own trade union had also been pursuing as a dual strategy of enforcing rights and raising awareness of working conditions:

> One of the things with isolated employment and individual employment is the fear factor but we've proved that people are prepared to come forward and we'll take it forward and challenge employers and I think these judgements are ground-breaking. (Trade Union: Senior Organiser)

Despite these efforts, it's clear that the gig economy poses challenges for trade unions in the UK and elsewhere. These relate not only to the difficulties of organising workers who are geographically dispersed and who are often working at different times of the day, but also to the fact that trade unions have been traditionally focused on workers who are clearly classified as working for one employer rather than self-employed. In some ways the challenges brought by the inadequacies of the traditional methods of organising workers have also brought opportunities for new movements and organisations to emerge. One example of this is the Independent Workers of Great Britain (IWGB), which describes itself as a new and dynamic union that focuses on marginalised workers and in particular those workers in the gig economy who are bicycle couriers or Uber drivers.[4] Interestingly, alongside the work it has done to further the rights of those working in the gig economy, it too has heralded its successes in campaigning for better conditions for other insecure workers, including a high-profile campaign to reverse the outsourcing of cleaning, security, reception and postal work staff at the University of London.[5] Therefore, what we can begin to observe is something of a parallel pathway of trade union activities of the larger, traditional and more established trade unions on the one hand and on the other hand those that are more grassroots organised and emergent, around the issue of insecurity (Dewhurst 2017).

Despite the challenges presented by organising gig economy workers, what our interviews revealed was a growing awareness that it was in fact possible to do so. Our findings are to some extent corroborated by recent developments in the gig economy in the UK, with examples of some agreements being reached at the firm level, for instance by the IWGB and an NHS contractor,[6] as well as between the GMB union and the distribution company Hermes.[7] These examples of company-level agreements have been pointed to by both unions as an illustration of the benefits of consistent campaign action and worker organisation. Across our interviews, however, there was an acknowledgement that organising in the gig economy did not come without its challenges. For example, one trade union officer we interviewed in the north of England lamented the fact that the sheer insecurity evident in the gig economy meant that workers were switching employers (and employment status) so frequently that it was making it incredibly difficult to organise them in comparison to other sections of the workforce:

> It is harder to organise gig workers because they don't tend to be in the same job for very long, they don't have the same opportunities to see union reps so it is much harder to organise them. (Trade Union: Policy and Campaigns Officer)

Nevertheless, although organising gig economy workers into trade unions is challenging, our interviews revealed that it was far from impossible, although doing so entailed accepting that there were risks and resources involved. One conclusion that could quickly be drawn from the interviews with trade unionists and the gig economy workers was that many issues could begin to be resolved by a change in their worker status. However, there was a degree of scepticism among trade unionists and some policymakers that simply moving workers from self-employed to employed status would be a silver bullet solution. This included one senior trade union official in Scotland who explained that, although he could understand the need to secure workers' rights in this way, it was not a panacea. He explained the danger that he perceived from such an approach:

> It sort of gets you off the hook from organising. (Trade Union: Policy Officer)

This sentiment was echoed by another trade union interviewee, this time a UK-level organiser who was equally sceptical and explained that, in his experience, just winning recognition as a worker (rather than being self-employed) should not be the end game for the trade union movement. Instead, the crucial element that would address issues of security and pay was the building of a collective voice among workers that could be translated into power (Lenaerts et al. 2018; Cant 2019; Vandaele 2021; Tassinari and Maccarrone 2020; Doherty and Franca 2020; Wood and Lehdonvirta 2021). The interviewee said he preferred to separate the issues of recruiting gig economy workers into trade unions on the one hand and organising them into a body of workers that could seek recognition from an employer for collective bargaining purposes on the other hand. He added that organising posed much more significant challenges than recruitment. The interviewee explained that he welcomed the recruitment of such workers to the trade union movement, but that it was quite another thing to ensure these same workers were being adequately serviced and the resources were in place to meet their needs.

> Unions have to say, 'Well we have a finite set of resources and what is the best use of those resources in organising?' ... If the union said we want to recruit all of the Uber drivers in a city, we could find a way to do that, but what are we recruiting them for? What are we recruiting them into? We're recruiting them into a union and we want to collectivise and turn that into a bit of power that they utilise to make changes to their relationship with Uber. (Trade Union: National Organiser)

The process of recruiting and organising workers was also explored by another interviewee from the trade union movement, who was organising young gig workers in the hospitality sector. The interviewee explained that he and his colleagues had focused efforts on organising those in the sector who have up until now been part of the workforce that are predominantly non-unionised. Doing so involved taking direct action against specific employers who had a track record of offering precarious and low-paid employment. The direct action initially took the form of workers actually shutting down the premises and doing so with high visibility in order to inflict as much reputational damage as possible on the company. Within weeks the company held discussions with the union and workers, and since then the union has established a dialogue with the employer and has increased union membership inside the company. Following the success of this direct-action repertoire, the union pursued a similar strategy with other employers that involved going through a series of stages:

> Mobilising advocacy, supporting workers, letting them know what their rights were, educating them about what they're entitled to as a form of empowerment. Moving into mobilising and taking direct action to show that workers can take direct action and win. The stage that we're in just now, we're still educating, we're still doing monthly training sessions for precarious workers. (Trade Union: Industrial Organiser)

What our interviews revealed, therefore, was that the challenges posed by the gig economy for trade unions to adequately recruit and organise workers were varied, but to successfully build the collective voice some had alluded to required information about the everyday working conditions of those in the gig economy. When exploring this issue with interviewees, it became clear that there was an increasing concern among policymakers but also in the trade union movement of a growing 'data deficit'. This was an issue explained to us by one interviewee as a key challenge when trying to understand how the gig economy may be impacting different sectors in the UK:

> You don't have good data on gig economy workers, so you've got somebody on zero hours contracts[,] which is not broken down by sector. (Trade Union: Policy Officer)

Another interviewee from the trade union movement echoed the concerns outlined above, highlighting an information asymmetry between workers and employers and raising a concern that part of the gap in information was caused by the trade unions falling behind in terms of understanding the technological changes taking place:

> There is a labour movement lag whenever there is a big change in the technology of work and there is always a relatively small group that's trying to convince the unions to catch up, so I wouldn't be that optimistic about the range of the campaigning tactics being deployed at the moment in the gig economy. (Trade Union: Campaigns Officer)

This concern of the trade unions in the UK struggling to keep pace with the technological changes taking place was echoed by another interviewee, who was involved in bringing together unions to collectively move forward the agenda on the future of employment. She explained that there was, in her view, a real need to engage with the challenges presented by the technology that was accompanying the growth of the gig economy because it had radically transformed some of the basics of work organisation. As an example, she pointed out that, in the past, the organisation of shift rotas would be done in the workplace. It was clear for all workers to see and was thus more transparent, but now,

> You get it through your company app which tells you you're going to be here today and there tomorrow, there's no instance where you can see what everybody else is doing, but also your manager isn't in charge of that as much as they would have been before. So actually, we need to argue with the algorithm and our arguing with the algorithm means understanding algorithms. (Labour Organisation: Director)

The interviewee added that these algorithms, coded by the employers, needed to be better understood by the trade union movement (see Lenaerts et al. 2018; Shapiro 2018; Wood et al. 2019; Tassinari and Maccarrone 2020; Duggan et al. 2020; Kellogg et al. 2020). However, it was clear from her words that there was a massive gap in terms of the available resources to meet these changes:

> I can name you three unions that have data analysts... That's the hidden problem. (Labour Organisation: Director)

5.   THE GIG ECONOMY AND ITS IMPLICATIONS FOR POLICYMAKING IN THE UK

The changing world of work has also provoked responses from governments across various contexts, and in this sense the UK is no different. Across our interviews it was clear that trade unions, labour organisations and even individual workers were actively making policymakers aware of the difficulties they were experiencing in the gig economy. Indeed, some interviewees, including some policymakers and trade unionists, insisted that there was now a need for a more radical approach to employment law to ensure better protections.

One way in which the UK Government has sought to better understand the development of the gig economy has been through commissioning the work of the Taylor Review of Modern Working Practices,[8] which had an objective of gathering inputs from various stakeholders across the UK, including voices from industry, employment law and the trade unions, to better grasp the impact of the changing world of work, including the gig economy. In the recommendations of the Taylor Review, a signpost is provided towards what is described as fair and decent work, that revolves around seven key points: the same basic principles applied to all forms of work in the UK economy; clarity in distinguishing workers from those who are truly self-employed; additional protections for those workers embedded in the gig economy; better corporate governance and employment relations within organisations; realistic ways for workers to strengthen their future employment prospects; a more proactive approach to worker health and wellbeing; and sector-specific strategies that seek to improve the pay and conditions of workers. In its response to the report, the UK Government has accepted some of the recommendations of the Taylor Review, such as bringing forward legislation that provides clarity in terms of employment status, but has chosen to take further consultation on the more specific recommendations that include developing a test for a new 'dependent contractor' status that provides legal clarification for those who are working in the gig economy, effectively distinguishing them from the standard understanding of what is a 'self-employed' worker (HM Government 2018). In practical terms this recommendation on employment status from the Taylor Review involved calling upon government to 'rename as "dependent contractors" the category of people who are eligible for worker rights but who are not employees'

(Taylor et al. 2017, p. 35). However, the response from a number of our interviewees to the Taylor report was one of disappointment at the recommendations made; and as we explored earlier in this chapter, unions organising in the gig economy have continued to pursue their own strategies and agreements with companies to secure employment status and rights for the workers they represent. Indeed, a key concern expressed by the trade union movement in the UK has been that creating a 'dependent contractor' status would generate more legal uncertainty for workers and weaken rights for some (Trades Union Congress 2018; see also Wood 2019).

One senior official in the trade union movement explained that there was at least one positive note that could be taken from the outcome of the Taylor Review, being that the issues around the gig economy were at least being highlighted. The interviewee indicated problems with the Taylor Review's recommendations, expressing scepticism about having a 'dependent contractor status' (see also Bales et al. 2018). She explained that there was a problem with the basic premise of the Taylor Review; that flexibility was a good thing; and that, although she agreed that people should have flexibility, this did not mean that their most basic employment rights should be denied. This sentiment was echoed by another interviewee from the trade union movement, a national level organiser, who was also very sceptical about the potential for policy change through initiatives such as the Taylor Review. Again, the interviewee expressed great caution about the idea posited by the review about the creation of a new category of worker:

> We didn't think it was a particularly rewarding set of recommendations. (Trade Union: National Organiser)

Moreover, the UK Parliament's Work and Pensions Committee has signalled a particular concern regarding the implications of the rolling out of the UK Government's flagship welfare reform, Universal Credit[9] (Dwyer and Wright 2014; Brewer and Handscomb 2020), for those working in the gig economy and called on the UK Government to ensure that Universal Credit is well positioned to meet the needs of gig economy workers. One specific recommendation in this area has been to request that the current Universal Credit system does not apply its Minimum Income Floor[10] to self-employed workers until the system can be recalibrated to respond to the insecurity of the gig economy. These issues are being considered not only by policymakers at the central government level in the House of

Commons, but also among representatives in the devolved institutions in the UK, with one such representative explaining:

> Universal Credit could potentially be an issue for gig economy workers because they may have a job for a month, three months, then nothing for a month. Is Universal Credit nimble enough to handle that? To plug the gaps and keep the wheels turning? I'm not sure, there is no evidence that it is. (Policymaker: Elected representative 1)

Echoing the views of other policymakers we interviewed, he emphasised that, although the previous system of support had been imperfect (working tax credits, a form of benefit that topped up low pay), in his view it was much more effective, whereas the new types of support were designed to encompass all types of workers under one umbrella (including the self-employed, the precariously employed and the unemployed). This was an issue that had been raised in his work on committees scrutinising social security policy as crucial in understanding the future of welfare in an age of the gig economy. Moreover, he added that, in his view, which was shared by colleagues in his party, the new system of Universal Credit had been designed in a way to deter workers from actually applying for support because they had not previously viewed themselves as requiring welfare support:

> You're seeing these people as claimants, but a lot of these people don't see themselves as claimants, they just see themselves as working and just need a bit of support, they don't see themselves as the same as someone who is unemployed. (Policymaker: Elected representative 1)

Other interviewees were equally concerned that, although they could understand the attraction of the gig economy for some workers in terms of the perceived flexibility it was offering, there was a sense that the pernicious effects of this type of work were affecting those sections of the population that were already suffering from poverty and inequality. This related to the types of contracts often found in the gig economy that were removing from workers very basic entitlements that had once been the minimum standards offered. This was exemplified by one politician we interviewed in Wales, who felt that the best way to begin eradicating insecure forms of employment was to remove them from use in the public sector and thus begin to roll back the culture of insecurity that had grown by setting standards in the public sector that could be replicated elsewhere in the economy:

> Our party has attempted to introduce restrictions on zero hours contracts. Particularly in the public sector – which should set a good example – if a government is opposed to these contracts, it should not allow their use. (Policymaker: Elected representative 2)

Another parliamentarian we interviewed, this time one who had a remit for social security policy, was particularly concerned at the extent to which gig economy type of work had become prevalent in the very poorest parts of his constituency and as a consequence poor quality work was being generated in already poor areas. This was a concern mirrored by another politician, who was explicit in her view that, as a result of the decline in incomes in her constituency, people were looking for more work to 'top-up' the poor incomes they were receiving for their existing employment and therefore were increasingly turning to jobs that she characterised as the gig economy.

Of course, although employment policy and significant parts of welfare policies remain the remit of the central UK Government, devolution in the UK means that alternative responses to the gig economy are possible. Policymaking areas such as education, skills development, employability, aspects of social security as well as fair work initiatives are tackled by devolved institutions and as such provide an arena for key actors, such as policymakers, employers and trade unions, to grapple with the implications of the gig economy. One example of that is Scotland, which is the constituent nation of the UK with the highest level of devolved powers. There, the Cabinet Secretary for the Economy, Jobs and Fair Work commissioned an Expert Advisory Panel on the Collaborative Economy to make recommendations to the Scottish Government on how Scotland can take advantage of the opportunities of the 'collaborative economy'[11] and meet the challenges that are brought by these new forms of employment for regulations and the economy. In their report, the panel recommended that the starting point for supporting Scottish workers in the gig economy should be to base the approach upon the principles that underpin the Scottish Government's Fair Work Framework, which stems from the Fair Work Convention established in April 2015 by the Scottish Government to promote better pay and conditions for workers in Scotland. The Fair Work Framework focuses upon developing effective voice among workers both individually and collectively, promoting fairer opportunities for workers to progress in their employment, ensuring greater security for workers specifically in terms of their contractual arrangements and finally, making sure that workers are treated respectfully regardless of their role or contractual status in the workplace. Of course, the implementation of these principles into policy is made somewhat problematic at the devolved level, given that employment policy remains reserved to central government at Westminster. Nevertheless, this has not prevented the Scottish Government from promoting a Living Wage among employers, and neither has it prevented the Expert Advisory Panel on the Collaborative Economy established by the Scottish Government from expressing concerns about the implications of the gig economy for the quality of work available in the country (Expert Advisory Panel on the Collaborative Economy 2018).

Of course, Scotland is not alone in having a devolved legislature in which issues such as the transformations in the labour market can be debated, even though the powers over employment policy also do not extend to the Welsh Assembly or the Welsh Government. Nevertheless, the challenges facing workers in a changing labour market have been the focus of attention of the Welsh Assembly's Equality, Local Government and Communities Committee, which gathered evidence from a variety of stakeholders across Wales and reported its findings in May 2018, thus filling a recognised gap in the knowledge base in Wales regarding the growth of non-standard forms of employment such as the gig economy (Henley and Lang 2017). As part of their findings, the Committee set out a number of recommendations (23 in total), ranging from the development of better indicators and benchmarks to encouraging employers to paying a living wage, as well as public procurement processes where the Committee recommended that the Welsh Government places requirements on any company receiving financial support from the Welsh Government to minimise the use of zero hours contracts and that companies should offer employees the option to move onto secure contracts after a set period of employment of around three months (Equality, Local Government and Communities Committee 2018).

In contrast to the devolved legislatures in Scotland and Wales, there has been little debate regarding the issue of the gig economy in the Northern Ireland Assembly, which can in part be attributed to the fact that the devolved assembly there remains suspended. However, this of course does not mean that workers in Northern Ireland have been immune from the changes brought by the gig economy, as evidenced by the Law Centre of Northern Ireland, which has also called for greater resources to be made available to HMRC (His Majesty's Revenue & Customs, the UK Government department responsible for the collection of taxes) to ensure companies are not using self-employment status to circumvent employment rights and for greater advice and support (in the form of making welfare benefits such as Universal Credit more responsive to workers' needs as well as maternity and childcare support for the self-employed) to be made available to those in the gig economy (Law Centre Northern Ireland 2017).

Across our interviews there were differing opinions with regard to the potential policy change that could be made via devolved government. For one interviewee, a policymaker based in Wales, there was little doubt that the situation regarding employment rights was one that needed to be challenged at the UK Government level because despite efforts in Wales to set out an agenda that could lead to better employment practices, there was a sense of there being only so much that policymakers could achieve at the devolved level and thus,

> If you were feeling negative you would say our ability is just to influence the edges and dance around the issue, the nub of the matter is one that can only be addressed through politics in Westminster. (Policymaker: Elected representative 3)

Another interviewee, a policymaker based in Scotland, explained that, in her view, the solution was to devolve more powers to the Scottish Parliament in order that a different path could be taken:

> If all employment rights and legislation was devolved we would be moving forward on this and dealing with these issues. That would ultimately for us be the answer. The Tories, whatever their motivation, it's not going to be strengthening workers' rights but meanwhile we will continue to push, as my colleagues are at Westminster, to make some improvements. (Policymaker: Elected representative 4)

Overall, there was a great deal of scepticism from most of the participants in our study that the legislation as it currently stands was fit for purpose in dealing with the issues generated by employment in the gig economy. Thus, our findings have opened a number of avenues for consideration in the design of future policy architecture underpinning the protections for workers in the UK gig economy. The issues raised across this chapter have implications not only for those policymakers whose remit encompasses areas of employment; they also extend to broader areas of social policy including the design and delivery of welfare systems as well as those responsible for enterprise and economic development.

6.   CONCLUSIONS

Overall, our findings provide a fresh, empirically informed insight into the implications of the gig economy for the future of worker protections in the UK. A number of key concerns emerged from our analysis. Firstly, those representing workers indicated that employment in the gig economy may exacerbate existing inequalities such as those experienced by women in the labour market. Also, there were concerns that the gig economy was at the forefront of loosening the relationship between workers and employers and consequently leading to a loss of control by the worker as well as diminished protections. However, despite the difficult environment in the UK described by some in the trade union movement, it was clear from our interviews that there are efforts being made to advance the protections of workers, sometimes directly in the workplace or via the courts. In fact, for the trade union movement, a major challenge going forward will be seeking to connect extant organising experience with innovative efforts to organise workers in the gig economy. What becomes clear in our findings from the UK is that both those representing workers and policymakers involved in the regulation of work in the gig economy will need to identify opportunities for more effective dialogue in order that protections can be developed to ensure that the disruption brought by technological change does not lead to widening inequalities or the proliferation of low-paid and insecure work. This is an issue not only for the workers themselves but also for the future of the welfare state in which they live and work. Given the nature of the online platforms, which have been driving a great deal of change, it may be incumbent upon researchers from across different scholarly disciplines engaged in the field of the future of work to appreciate the need for realising opportunities for interdisciplinary research agendas with colleagues in data science focused disciplines.

NOTES

1.   http://www.legislation.gov.uk/ukpga/2016/15/contents/enacted   (accessed on 9 June 2021).
2.   https://www.gmb.org.uk/news/uber-finally-does-right-thing-after-gmb -wins-four-court-battles (accessed on 9 June 2021).

3. https://www.theguardian.com/business/2021/may/26/uber-agrees-historic
-deal-allowing-drivers-to-join-gmb-union (accessed on 9 June 2021).
4. https://iwgb.org.uk/page/about/about (accessed on 14 January 2020).
5. https://iwgb.org.uk/en/post/5b5f446a12f0a/iwgb-campaign-wins-major
-conce (accessed on 14 January 2020).
6. https://www.iwgb.org.uk/post/5d0c844ddd65b/iwgb-reaches-historic-pay
-deal (accessed on 14 January 2020).
7. https://www.gmb.org.uk/news/hermes-gmb-groundbreaking-gig-economy
-deal (accessed on 14 January 2020).
8. https://www.thersa.org/globalassets/pdfs/reports/good-work-taylor-review
-into-modern-working-practices.pdf (accessed on 9 June 2021).
9. Universal Credit is a welfare benefit that is being introduced as an all-encompassing form of welfare payment that replaces existing benefits such as Housing Benefit; Child Tax Credit; Income Support; Working Tax Credit; Income-based Jobseeker's Allowance; and Income-related Employment and Support Allowance.
10. Universal Credit payments are to be based on the assumption that someone is earning a certain amount through self-employment, even if they don't actually earn this much. This assumed amount is called the 'minimum income floor'.
11. Used here as a synonym for the gig economy.

REFERENCES

Aslam, Y., and J. Woodcock (2020). 'A history of Uber organizing in the UK', *South Atlantic Quarterly*, **119** (2), 412–421.
Autor, D. H. (2001). 'Wiring the labor market', *The Journal of Economic Perspectives*, **15** (1), 25–40.
Bales, K., A. Bogg and T. Novitz (2018). '"Voice" and "choice" in modern working practices: Problems with the Taylor Review', *Industrial Law Journal*, **47** (1), 46–75.
Brewer, M., and K. Handscomb (2020). 'This time is different – Universal Credit's first recession', Resolution Foundation, London.
Bryman, A. (2016). *Social Research Methods*. Oxford: Oxford University Press.
Burns, E. (2010). 'Developing email interview practices in qualitative research', *Sociological Research Online*, **15** (4), 24–35.
Cant, C. (2019). *Riding for Deliveroo: Resistance in the New Economy*. Cambridge: Polity.
Crouch, C. (2019). *Will the Gig Economy Prevail?* Cambridge: Polity.
Department for Business, Energy & Industrial Strategy (2018). *The Characteristics of Individuals in the Gig Economy*. London: BEIS.
Department for Business, Energy & Industrial Strategy (2021). *Trade Union Membership: Statistical Bulletin*, 27 May. London: BEIS.
Dewhurst, M. (2017). 'We are not entrepreneurs', in M. Graham and J. Shaw (eds), *Towards a Fairer Gig Economy*, pp. 20–23. Manchester: Meatspace Press.

Doherty, M., and V. Franca (2020). 'Solving the "gig-saw"? Collective rights and platform work', *Industrial Law Journal*, **49** (3), 352–376.

Duggan, J., U. Sherman, R. Carbery and A. McDonnell (2020). 'Algorithmic management and app-work in the gig economy: A research agenda for employment relations and HRM', *Human Resource Management Journal*, **30** (1), 114–132.

Dwyer, P., and S. Wright (2014). 'Universal Credit, ubiquitous conditionality and its implications for social citizenship', *The Journal of Poverty and Social Justice*, **22** (1), 27–35.

Equality, Local Government and Communities Committee (2018). *Making the Economy Work for People on Low Incomes*. National Assembly for Wales, Cardiff Bay.

Expert Advisory Panel on the Collaborative Economy (2018). *Final Report*. Scottish Government, Edinburgh.

Ford, M., and T. Novitz (2016). 'Legislating for control: The Trade Union Act 2016', *Industrial Law Journal*, **45** (3), 277–298.

Goos, M., and A. Manning (2007). 'Lousy and lovely jobs: The rising polarization of work in Britain', *The Review of Economics and Statistics*, **89** (1), 118–133.

Goos, M., A. Manning and A. Salomons (2014). 'Explaining job polarization: Routine-biased technological change and offshoring', *American Economic Review*, **104** (8), 2509–2526.

Graham, M., I. Hjorth and V. Lehdonvirta (2017). 'Digital labour and development: Impacts of global digital labour platforms and the gig economy on worker livelihoods', *Transfer: European Review of Labour and Research*, **23** (2), 135–162.

Gregg, P., and L. Gardiner (2015). *A Steady Job? The UK's Record on Labour Market Security and Stability since the Millennium*. Resolution Foundation, London.

Hall, P. A., and D. Soskice (eds) (2001). *Varieties of Capitalism: The Institutional Foundations of Comparative Advantage*. Oxford: Oxford University Press.

Henley, M., and M. Lang (2017). 'Self-employment in Wales: Micro-business activity or the rise of the gig economy?', *Welsh Economic Review*, **25**, 9–17.

HM Government (2018). *Good Work. A Response to the Taylor Review of Modern Working Practices*. London: HM Government.

Huws, U. (2014). *Labor in the Global Digital Economy: The Cybertariat Comes of Age*. New York, NY: New York University Press.

Hyman, R., and R. Gumbrell-McCormick (2017). 'Resisting labour market insecurity: Old and new actors, rivals or allies?', *Journal of Industrial Relations*, **59** (4), 538–561.

Kellogg, K. C., M. A. Valentine and A. Christin (2020). 'Algorithms at work: The new contested terrain of control', *Academy of Management Annals*, **14** (1), 366–410.

Law Centre Northern Ireland (2017). *Work & Pensions Committee: Self Employment and the Gig Economy*. Belfast: Law Centre NI.

Lenaerts, K., Z. Kilhoffer and M. Akgüç (2018). 'Traditional and new forms of organisation and representation in the platform economy', *Work Organisation, Labour and Globalisation*, **12** (2), 60–78.

McGuinness, F. (2017). *Income Inequality in the UK*. Briefing Paper, House of Commons Library, Number 7484.

Office for National Statistics (2018). *Trends in Self-employment in the UK: Analysing the Characteristics, Income and Wealth of the Self-employed*. February, London.

Office for National Statistics (2019). *Labour Disputes in the UK: 2018*. London: ONS.

Plunkett, J., and J. P. Pessoa (2013). *A Polarising Crisis? The Changing Shape of the UK and US Labour Markets from 2008 to 2012*. Resolution Foundation. London.

Resolution Foundation (2017). *Freshly Squeezed: Autumn Budget 2017 Response*. London.

Scholz, T. (ed.) (2012). *Digital Labor: The Internet as Playground and Factory*. Abingdon, Oxon: Routledge.

Scholz, T. (2017). *Uberworked and Underpaid: How Workers are Disrupting the Digital Economy*. Cambridge: Polity.

Shapiro, A. (2018). 'Between autonomy and control: Strategies of arbitrage in the "on-demand" economy', *New Media & Society*, **20** (8), 2954–2971.

Shead, S. (2017). 'Uber loses appeal in landmark UK case over its drivers' employment rights', at http://uk.businessinsider.com/uk-judge-rejects-uber-appeal-over-driver-employment-rights-2017-11 (accessed: 5 January 2020).

Standing, G. (2011). *The Precariat: The Dangerous New Class*. London: Bloomsbury Academic.

Tassinari, A., and V. Maccarrone (2020). 'Riders on the storm: Workplace solidarity among gig economy couriers in Italy and the UK', *Work, Employment and Society*, **34** (1), 35–54.

Taylor, M., G. Marsh, D. Nicol and P. Broadbent (2017). *Good Work: The Taylor Review of Modern Working Practices*. London: Department for Business, Energy & Industrial Strategy.

Thompson, E. P. (1967). 'Time, work-discipline, and industrial capitalism', *Past & Present*, **38**, 56–97.

Trades Union Congress (2018). *Taylor Review: Employment Status. TUC Response to the BEIS/HMT/HMRC Consultation*. London: Trades Union Congress.

Vandaele, K. (2021). 'Collective Resistance and Organizational Creativity amongst Europe's Platform Workers: A New Power in the Labour Movement?', in J. Haidar and M. Keune (eds), *Work and Labour Relations in Global Platform Capitalism*, pp. 205–234. Cheltenham, UK and Northampton, MA, USA: Edward Elgar Publishing; and Geneva: ILO.

Wood, A. (2019). 'The Taylor Review: Understanding the gig economy, dependency and the complexities of control', *New Technology, Work and Employment*, **34** (2), 1–9.

Wood, A., and V. Lehdonvirta (2021). 'Antagonism beyond employment: how the "subordinated agency" of labour platforms generates conflict in the remote gig economy', *Socio-Economic Review*, **19** (4), 1–44.

Wood, A. J., M. Graham, V. Lehdonvirta and I. Hjorth (2019). 'Good gig, bad gig: Autonomy and algorithmic control in the global gig economy', *Work, Employment and Society*, **33** (1), 56–75.

6. Regulating digital crowdwork and the need for global responses

Maria Mexi and Konstantinos Papadakis

1. INTRODUCTION

The growth of the digital platform economy is one of the most important developments transforming the world of work. The very technology powering the platform economy has led to an increasing global division of labour via web-based labour platforms where work is outsourced by businesses and other clients through an open call to a globally geographically dispersed crowd, also known as online web-based platforms termed "crowdwork", the main focus of the present chapter.[1] On the positive side, web-based platforms and crowdwork are enabling a global mobility of virtual labour, having the potential to help jobseekers from low- and middle-income countries (including women, people with disabilities, youth and migrant workers) enter new labour markets, often in wealthier economies, that were previously out of reach due to migration barriers (ILO 2021). It can thus offer income, skill utilization and enhancement, international exposure, assured payments for work undertaken, and "gainful employment opportunities for low-skilled developing country workers who are unemployed, underemployed or in the informal sector" (Berg et al. 2018, p. 88). On the negative side, platform-enabled crowdwork means that firms can gain access to a global diverse pool of workers from which to recruit at low cost (ILO 2021). As a result, some point to a race to the bottom on wages and workers' rights, related to geographical differences in skills and labour costs. Overall, digital labour platforms could likewise contribute to (further) fragmentation of production into "business units" across national jurisdictions, eroding employment relationships and protections. International regulation of labour platforms thus appears necessary.

Against this background, the purpose of this chapter is to shed light upon the following fundamental question: How can we address inequities fostered by cross-border platform crowdwork and create spaces for promoting democratic governance and participation of all actors in the crowdwork platform industry, enabling them to work together to ensure level playing-field rules against social dumping, while reverting to a possible degradation in basic labour standards? Responding to this challenge, we argue that inspiration can be taken from voluntary cross-border social dialogue initiatives and agreements, such as the transnational company agreements (TCAs). Such agreements include international framework agreements between Global Union Federations and multinational enterprises, and European framework agreements between multinational enterprises and European trade union federations and/or European Works Councils. As explained below, due to their qualities and outcomes (enhancing reflexivity in the management of labour processes and economic activities that are no longer territorially limited while promoting labour standards) tested in a global supply chain environment, both cross-border social dialogue and TCAs can give rise to new opportunities for facilitating dialogue, and promoting coordination between digital labour platforms and workers responding to the emergence of a global or "planetary labour market" (Graham and Anwar 2019).

The next sections explore a number of key challenges pertaining to the governance of the global platform economy and work, and particularly regulating platform work across multiple jurisdictions. This analysis takes place within a growing literature examining the particular role assigned to social dialogue actors and initiatives beyond national borders, including value chains. For those studying the platform economy, there are key insights to be drawn from the ways in which cross-border social dialogue actors and initiatives not only shape transnational labour (self-) regulation, but also contribute to the emergence of global industrial relations frameworks in the absence of a single multilateral framework regulating social and labour rights (Papadakis 2011b).

2.   PLATFORM-BASED WORK: STRUCTURE AND KEY CHALLENGES

The significance and rapid growth of the platform economy is apparent. Overall, driven by technological advances and increased online connectivity, these platforms are increasingly reshaping the world of work at both national and global levels. The total revenue of Upwork, one of the biggest digital labour platforms globally, increased from US$164 million in 2016 to US$301 million in 2019 (ILO 2021). Concurrently, the taxi service Uber was founded in 2009 and quickly grew to become the world's fastest-growing start-up, with a market valuation of over US$90 billion in 2021.[2] As the platform economy grows, there are new challenges that arise. A key challenge is related to the fact that the rise of the platform economy comes with heightened complexity, which characterizes platform work as well as the digital labour platforms' operation and outreach.

In seeking to understand its complexity, scholars have developed several terminologies and taxonomies. For instance, Heeks (2016, 2017), drawing on definitions provided by Horton (2010), Lehdonvirta et al. (2014) and Graham et al. (2017), describes that the platform economy encompasses "online labour" from a work and labour perspective with "online outsourcing" from a client-side focus. In this framework, "online labour" is defined as contingent (task- or project-based) intangible work delivered digitally and done for money, organized via online outsourcing platforms that are marketplaces bringing together buyers and sellers. Schmidt (2017) then suggests a taxonomy of online labour divided into two types: (a) "crowdwork", where tasks allocated via platforms (for instance, Amazon Mechanical Turk) are not given to a specific individual but are further sub-divided into small units of piecemeal task in the form of microwork and contest-based (many individuals compete for the task but only one result is used and paid for, for example, 99designs); and (b) "online freelancing", where a more substantial task (for instance, software development, translation, administrative support, etc.) is assigned to an identified individual via platforms such as Upwork or Freelancer (see also Agrawal et al. 2013; Margaryan 2016).

The reviewed literature further offers different conceptions of crowdwork or crowd employment. While some authors (Flecker et al. 2017; Leimeister et al. 2016) understand crowdwork as a type of labour that is intermediated by a platform and carried out remotely involving platform-mediated remote service provision (Kenney and Zysman 2015), others (Huws et al. 2016; Drahokoupil and Fabo 2016, 2017; Codagnone et al. 2016) define crowdwork as paid work that is intermediated and organized by an online platform whatever the location of provision (online or local). In addition, Durward et al. (2016) develop one of the most extensive frameworks for crowdsourcing and crowdwork. They distinguish between internal and external crowdwork, depending on whether the crowd is employed by the platform or not. External crowdwork is further broken down into a number of categories such as microtask platforms (which involve simplified and repetitive tasks) and marketplace platforms (where tasks take longer to complete and are predominantly complex). There are further classifications. For instance, other scholars adopt a double distinction of platform work, highlighting space-less or space-full attributes of work. More specifically, Codagnone et al. (2016) differentiate between (a) platforms coordinating local services (so-called mobile labour markets, MLMs) that basically involve local markets and include taxi services, household services, home repair, pet-sitting services, legal services, etc., and (b) platforms intermediating online services (so-called online labour markets, OLMs) and splitting jobs into micro-tasks or Human Intelligence Tasks, HITs (as named on Amazon Mechanical Turk). OLMs are global markets and entail a global division of tasks (Eurofound 2015). In this context, Horton (2010) and Degryse (2016) notice that OLMs enable virtual labour migration. This understanding is crucial in terms of grappling with the ways in which the platform economy has contributed to the gradual emergence of a global workforce dispersed across multiple jurisdictions around the globe, and enabling businesses to source talent globally (Prassl 2018; Mexi 2020; ILO 2021).

Crowdwork platforms are becoming more and more globalized, involving workers that perform outsourced tasks online in different parts of the world. This, in effect, is increasingly challenging states' regulatory and enforcement capacity, adding multiple levels of complexity. In particular, while the platform businesses that direct workers to deliver local services (for example, Uber, Lyft, Deliveroo) operate within the remit of the regulations of the national/local jurisdiction(s), the global nature of work outsourced via crowdwork platforms (for instance, Upwork, Amazon Mechanical Turk, etc.) to a geographically dispersed crowd or to selected individuals (freelancers) raises a number of complex questions for both national and international regulatory systems. It is important to keep the dimension of platforms' global operation in mind as well as its growth rate. Figure 6.1 below provides evidence on the increase in crowdwork facilitated by online web-based platforms between 2017 and 2021. As Stephany et al. (2021, p. 3) note: "In early 2021, roughly 90% more projects were demanded via online freelance platforms than in mid-2016 when the Online Labour Index started [Figure 6.1]. This equals an annual growth rate of 10%, which is significantly higher than changes in national (on-site) labour markets, which have plummeted in many countries as a result of the Covid-19 pandemic."

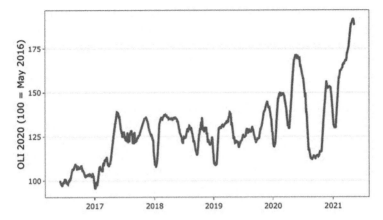

Figure 6.1   *Online global labour demand of projects via online web-based platforms, 2017–2021*

Source: OLI (2020) and Stephany et al. (2021).

The distinguishing characteristic of crowdwork facilitated by web-based platforms is that online connectivity enables work to be performed remotely from any location and to be outsourced across borders. Hence, web-based crowdworkers are spread across the globe, and many live in different countries from where the platform business or the crowdsourcer client is based.

The global span of much work performed via crowdwork platforms means that it tends to be done outside of the purview of national labour standards and rules in either their employers' home countries or the platform worker's home country. Hence, what is fundamentally different in such forms of platform economy from pre-existing work patterns is the globally dispersed nature of crowdwork that raises crucial questions about the exact boundary position of possible regulatory – including governance – interventions. How would regulations apply to work that is performed in Switzerland but delivered online via a digital platform registered in the US to a crowdsourcer client located in yet another country (or sometimes involving multiple platforms and clients in different countries)? And even if a specific regulatory model is designed to be relevant, how can it be enforced across borders? These characteristics raise issues of regulation and enforcement that cannot be effectively tackled solely through interventions addressing working conditions at the national level.

The ILO's Global Commission on the Future of Work has proposed an international governance system for crowdwork platforms, requiring platforms (and their clients) to respect certain minimum rights and protections, on the model of the ILO's Maritime Labour Convention for the global shipping industry. At the heart of such proposal is the recognition that, if left largely unregulated and unchecked, the platform economy threatens to undermine dignified working conditions and quality jobs in both low- and high-income countries as well as erode employment relationships and protections (ILO 2019c). If these regulatory gaps are not addressed, even if crowdwork is not widespread among most sectors, some of its effects are likely to "spill over" into more traditional types of work, spreading and intensifying precarity at work (Berg 2019; Garben 2019).

3. REGULATING PLATFORM-BASED CROWDWORK ACROSS MULTIPLE JURISDICTIONS: KEY ASPECTS AND OPPORTUNITIES

In recent years, discussion has focused on the challenges of (top-down) regulation in the platform economy and on the suitability of an international regulatory framework for cross-border, web-based, crowdwork (Berg et al. 2019), yet without advancing concrete proposals on the tools required for such a framework. We argue that self-regulation, with cross-border social dialogue at its core, could be a realistic option.

Cross-border social dialogue refers to,

> [the] dialogue developed between or among governments, workers and employers or their representatives beyond national borders for the purpose of promoting decent work and sound industrial relations. Such dialogue may focus on the opportunities and challenges associated with a country, economic sector, enterprise, region or group of countries. It may take place within ad hoc or institutionalized fora, mechanisms involving two or more parties or as a result of public and/or private (self-)regulatory initiatives developed in the context of a dynamic of economic integration and the organization of production along increasingly complex global supply chains. (ILO 2019a, p. 2)

Cross-border social dialogue is not new. Since 1919, "the practice of bringing representatives of governments, employers and workers together at the international level to seek consensus-driven solutions to socio-economic issues has become a key feature of the multilateral system, originating with the creation of the ILO. Today, cross-border social dialogue continues in the executive bodies of the ILO, in global, sectoral and regional meetings" (ILO 2019a, p. 3 – see also Bonvin 1998). Such meetings and their outcomes (such as declarations, codes of practice, guidelines, conclusions and points of consensus) are intended to shape attitudes, policies and regulations at all levels. The agenda of ILO meetings is itself the outcome of cross-border social dialogue. "The spaces for cross-border social dialogue have multiplied over the past century in response to deepening globalization and regional integration" (ibid.). Cross-border social dialogue is emerging and consolidating in both public and private governance spaces, as globalization continues to raise three broad challenges: (a) an ineffective implementation and enforcement of national labour laws and regulations in many countries, which prevents the improvement of working conditions that could result from increased participation in global supply chains; (b) the relocation of some particularly labour-intensive processes from developed economies to countries with lower labour costs, which intensifies the fear, or reality, of social dumping; and (c) the widening income inequality and a declining wage share in many countries' gross domestic product, related in large part to the erosion of collective bargaining and the deteriorated climate of industrial relations in these countries (ILO 2018b).

In the last five years, the ILO has firmly placed the question of cross-border social dialogue at the top of its policy agenda. In 2016, following a general discussion on decent work in global supply chains in the International Labour Conference, the Organization was called upon to promote cross-border social dialogue, for instance by offering assistance, "upon joint request", in the design and follow-up phase of international framework agreements, "including monitoring, mediation and dispute settlement where appropriate." Importantly, in early 2019 the ILO's mandate in the area of cross-border social dialogue was further consolidated on the occasion of a tripartite Meeting of Experts, which reviewed existing standards, practices developed at various levels, and the instances that pave the way for cross-border social dialogue and agreements. The "Conclusions" reached by the experts, approved by the ILO Governing Body in November 2019, called upon ILO Member States to create an enabling environment for cross-border social dialogue by, building the capacity of labour administrations and labour inspectorates in relation to cross-border social dialogue; adopting national policies and regulations that are conducive to cross-border social dialogue; and promoting effective linkages between different forms and levels of social dialogue and strengthen their complementarity (ILO 2019b).

As a result of campaigns by international trade unions or at the initiative of big enterprises with a deeply rooted social dialogue culture, some multinational enterprises and global unions have since the late 1980s increased cross-border collaboration through negotiating agreements such as the TCAs. Above all, TCAs are joint management-union initiatives, which entail corporate recognition and the participation of Global Union Federations (or European Industry Federations) as core partners in the negotiation and implementation of agreements. They often include direct references to ILO conventions, and they focus predominantly on creating conditions conducive to the organization of workers, trade union activity and collective bargaining down the value chains of multinational enterprises concerned by these agreements. TCAs, particularly international framework agreements, increasingly refer to global supply chains and contain provisions stating that subcontractors and suppliers must comply with the international framework agreements (ILO 2018a). Transnational company agreements have a proven record of helping to solve local conflicts down the value chains of multinational enterprises through joint grievance mechanisms established by the agreements; they trigger trade union campaigns that may lead to an increase in the unionization rate; and they boost collective bargaining for trade union recognition and the improvement of working conditions (Papadakis 2011b).

Overall, transnational company agreements aim to establish ongoing relationships between multinational enterprises and Global Union Federations, for the benefit of both parties. They are intended to promote principles of labour relations and conditions of work – notably in the area of freedom of association and collective bargaining – and to organize a common labour relations framework at cross-border level. The emergence of this form of transnational private labour regulation via transnational company agreements has proved crucial in addressing governance gaps in increasingly complex value chains that span across borders. Between 1988 and early 2022, over 400 TCAs were signed by some 200 enterprises, mostly European-based multinational enterprises. Several policy documents adopted by the ILO have depicted evidence of the positive impact that transnational company agreements may have on the improvement of, and compliance with, labour standards within multinational enterprises operations and supply chains.[3]

It should be noted that TCAs cannot be defined as "collective agreements" of a global scale, according to the ILO definition of collective agreements (Papadakis et al. 2008), but they have proved to be an important tool for managing complexity, as in the case of global supply chains, and to moderate the effects of a race to the bottom on wages and workers' rights across different economic geographies of work. In this respect, TCAs can provide useful lessons for agreeing to adopt analogous instruments so as to fit the complexity of the global platform economy that links virtual crowdwork labour with a net of (often) multiple entities,[4] physical locations and different regulatory domains where work is performed, mediated and delivered. At the same time, the inherent adaptability and reflexivity of social dialogue taking place at the cross-border level presents a unique opportunity to ensure decent working conditions for all platform workers in the platform economy and to accommodate "voice", in Hirschman's sense, by transforming relations between platform companies and platform workers through the creation of spaces for democratic consultation.

However, cross-border social dialogue requires strong structures of collective representation and voice. And efforts are emerging to organize platform workers, with the support of trade unions or through new ways of collective mobilization. The following examples are highlighted in the literature.[5] In the United States, the team behind Turkopticon, an online community of Amazon Mechanical Turk platform workers, created a web platform called "Dynamo" that focuses specifically on building collective action (Bergvall, Kåreborn and Howcroft 2014). The Dynamo platform set forward a campaign of sustained collective action around the publication of guidelines for academic requesters using Mechanical Turk, addressing matters such as fair pay (Salehi et al. 2015). Besides, workers with traditional employer–employee relationships as well as platform workers are using platforms like Coworker.org to test early forms of digital employee network-building via user-generated petition campaigns. Many platform workers also assemble on Facebook and WhatsApp groups, sub-Reddits, and other digital points of assembly to share experiences, chat, complain and exchange information, building solidarities in hyper-local contexts (Forlivesi 2018). Alternative digitally enabled mobilizations have included the organization of strikes and boycotts in the delivery sector involving delivery workers logging out en masse from apps that allocate work shifts (Forsyth 2019). Hence, new forms of virtual mobilization are emerging; and although it remains an open question what their impact will be, there is something on which to build.

A more concerted effort is needed to support agency among platform workers, especially those at opposite ends of the globe, and to effectively include them in global labour standards. As discussed above, the pressure on wages of platform workers is high and competition is global. Solutions promoting collective mobilization and representation at global level are being explored, through which trade unions would engage with platforms to ensure that basic working conditions are consistently applied. To achieve this aim, apart from cooperation between unions in the global North and South, one approach would be to effectively include platform workers in existing trade unions, including global structures of worker representation (and, subsequently, social dialogue, including collective bargaining, at national or cross-border level; see Papadakis 2021). Another approach would be to establish joint platforms or alliances between different national unions or Global Union Federations (given the cross-sectoral dimension of crowdwork) intended to collectively organize and represent platform workers at national or cross-border level. Their aim would be to take into account the full scope of platform work and to capture the particular issues pertaining to interest representation that are unique for platform workers in specific industries and sectors based on earlier cross-border union collaboration (see Davies et al. 2011). One such example is the formation of the Transnational Federation of Couriers in October 2018, which represents platform workers across Europe and aims to improve working conditions in the platform economy across Europe.[6]

On the other side of the table, the digital platforms' side, the situation appears more complex, to the extent that only few collective organizations representing platforms exist to date. This seems to be related to the fact that digital labour platforms rarely see themselves as an employer. Indeed, most platforms perceive themselves as intermediaries. And while at the national level there is evidence on the willingness of digital labour platforms to act as employers' organizations or opt for representation through existing employers' organizations (see Table 6.1 below), there is no such evidence at the cross-border level.

The *status quo* is likely to change, though under certain conditions. As described more analytically elsewhere (Papadakis 2021), awareness of a need to conduct social dialogue – and reach tangible, bipartite, cross-border agreements – has largely relied on interconnected factors or incentives that could also be important for cross-border social dialogue involving digital labour platforms.

More specifically, pressure from "within", for instance, can play a role in workers' organizing, mobilizing and protesting, on the one hand, meeting a responsible corporate culture combined with a wish to prevent competitors from freeriding on the other. The maritime sector is a good example. It is the only sector covered by a fully fledged global collective agreement between international social partners – that is, shipowners represented by the International Maritime Employers' Council and seafarer representatives from across the globe represented by the International Transport Workers' Federation. The agreement regulates wages and other terms and conditions of work, including maternity protection, and has strong links with ILO standards. First negotiated in 2001, it has been regularly re-negotiated and updated under the auspices of what is now known as the (bipartite) International Bargaining Forum. Once the agreement is negotiated, the unions affiliated to the International Transport Workers' Federation begin local negotiations with companies in their country, resulting in national industry or company level collective agreements. While entitlements may vary slightly, all agreements must be within the framework of the global agreement agreed for the period. This global sectoral agreement finds its roots in the 1990s, which witnessed a progressive consensus among global social partners on the need to launch collective bargaining of a global scale in the maritime sector. Shipowners had previously formed the International Maritime Employers' Council for the purpose of negotiating a global industry pay agreement with the International Transport Workers' Federation, for seafarers working aboard "flag of convenience" ships – that is, ships profiting from a global maritime regime that allows shipowners to avoid unions and regulation by flagging their vessels in countries with weak regulatory systems. The "flag of convenience" was the key federating theme (or a "common enemy") between a majority of shipowners and seafarers. For the former, "flag of convenience" represented the epitome of unfair competition by rogue shipowners who were disrespectful of minimum pay and of the regulations concerning working conditions and occupational safety and health. For the latter, "flag of convenience" led not only to bad working and pay conditions and dangerous workplaces, but also to a downward spiral of social conditions prevailing mostly in the European maritime sector (Papadakis 2021).

Another factor that could trigger the development of cross-border social dialogue in the platform economy is pressure from "outside", which can stem from public awareness of the suboptimal working conditions of platform workers (on the increase due to the pandemic). Such external pressure could incentivize all sides to enter dialogue, especially in the presence of proactive social-partner organizations and civil society more broadly.

In this regard, the international campaign against child labour in the Sialkot soccer ball industry in Pakistan is an early example. This case was one of the first illustrations on how a trade-union-initiated (International Confederation of Free Trade Unions) awareness-raising campaign, using "name and shame" techniques of transnational advocacy networks (Keck and Sikkink 1998), can effectively pressure multinational enterprises and governments to be part of cross-border social dialogue processes and reach agreement on workers' rights. This campaign paved the way, in 1997, to the signature of the Atlanta Agreement, which committed the Sialkot Chamber of Commerce and Industry, the ILO and UNICEF, various Pakistani NGOs, the Government of Pakistan, Save the Children UK, and the World Federation of Sporting Goods Industry to work together on eliminating child labour from the football industry in Sialkot (Baccaro 2001). A more recent example has been the Accord on Fire and Building Safety signed in the aftermath of the Rana Plaza building collapse in 2013. The Accord (renewed twice since then) is the first binding agreement gathering together more than 200 global brands and retailers in the textile and apparel sector, two global union federations (UNI Global and IndustriALL) and eight Bangladeshi unions in order to work towards a safe and healthy garment and textile industry in Bangladesh (ILO 2019a).

The "shadow of public regulation" is another crucial element compelling actors to join cross-border social dialogue and agreements on work-related matters. For instance, the shadow of regulation functioned as a key trigger for the first TCAs, signed in the late 1980s. Until then no multinational enterprise had accepted the invitation to join in cross-border social dialogue, in spite of pressures from the international labour movement throughout the 1970s demanding transnational collective bargaining on wages and working conditions, the outcomes of which would become globally binding. Their stance started to change in the 1980s only when a proactive European Commission announced its intention to make negotiations with the central management of multinational enterprises compulsory in the case of a planned transnational restructuring – the so-called "Vredeling directive" draft of 1980 (da Costa and Rehfeldt 2008).

It was in the context of this debate that the first multinational enterprises accepted to participate in cross-border social dialogue and, indeed, to sign the first transnational company agreements (Rehfeldt 1998). This "regulatory threat" by the European Commission and the ensuing TCAs triggered several developments. Among other things, the central management of multinational enterprises formally recognized global trade union organizations as their "bargaining" partners at the cross-border level (Papadakis 2011a). Multinationals also allowed the creation of transnational employee representation with information rights (so-called "European group committees" and similar bodies). The latter paved the way for the adoption of the European Works Council directive in 1994, which formalized transnational employee representation in multinational enterprises operating in the European Union territory. More recently, the emerging mandatory "due diligence" regulation targeting global supply chains (Bright 2021), in many European countries but also at the European Union level,[7] seem to present a contemporary version of the "Vredeling directive", in that it is likely to produce "boosting" effects as far as transnational social dialogue is concerned. With regard to crowdwork platforms, the European Commission proposed in December 2021 a set of measures (including a directive) to improve the working conditions in platform work and ensure that "people working through digital labour platforms can enjoy the labour rights and social benefits they are entitled to."[8] The shadow of European regulation seems to have operated once again as an incentive, for the first cross-border social dialogue involving Uber, to take root (see below).

The aspect of regulating the platform economy via social dialogue increasingly gains importance. In the literature, social dialogue is described as an instrument for providing a voice to key stakeholders by opening venues and levels for participation in decision-making processes (Papadakis 2006; Didry and Jobert 2011; de Munck et al. 2012; ILO 2013a; ILO 2013b). As part of this attribute, social dialogue is especially competent in shaping new win–win solutions and tackling collective action problems. By promoting consensus-building on substantive norms and "ownership" of the policies adopted by the parties concerned, it neutralizes and rectifies imbalances in a faster and more flexible and tailored way than regulatory interventions and individual litigation. Social dialogue thus can function as an effective regulatory alternative. As the need to bring the platform economy into the scope of social dialogue – and self-regulation – is becoming more and more pertinent, there are visible signs, as Table 6.1 shows, that platform economy actors are (hesitantly) beginning to engage in bipartite and other forms of dialogue regarding especially on-location platform work taking place at national level (Mexi

2019). These developments associated with social dialogue initiatives taken by platform economy actors at national level might incentivize cross-border social dialogue and their effects are likely to "spill over" into the global level.

If anything, these experiences are informative in the sense they illustrate that "systems are able to adjust to cover different and new forms of work" (OECD 2019). Several factors (often interconnected) contribute to such adjustments through social dialogue.

One strong factor determining propensity for social dialogue and voluntary agreements at national level (for example, collective agreements, codes of conduct, collaboration memoranda, etc.) lies in the tendency of some platforms to act in a socially responsible way and present themselves as a fair option. In other words, it can stem from the platform's set of strategic considerations such as attracting socially sensitive customers or skilled workers in tight employment markets. This may also come as a response to national (self-)regulation frameworks aimed at making crowdwork fairer. For instance, in France, a legal provision that encourages platforms to publish "social responsibility charters" online to be appendixed to workers' contracts is under exploration. Such charters would state the platforms' policy on work-related issues such as the prevention of occupational risks, professional development, measures to guarantee a decent income to workers, and rules framing the communication of changes in working conditions. Similarly in Germany, at the initiative of the *Deutscher Crowdsourcing Verband*, a code of conduct has been established and signed in 2017 by eight German-based platforms in collaboration with IG Metall. The platforms united in the *Verband* also collaborated with IG Metall in the establishment of an Ombudsman's Office that serves as a dispute settlement mechanism. Another significant initiative is the "Frankfurt Declaration on Platform-Based Work" in 2016,[9] endorsed by trade unions across Europe and the United States. The key merit of the Declaration is that of clearly strengthening a number of fundamental principles related to platform work at transnational level that may need to be fuelled down to national contexts.

The multiple legal cases and emerging laws aimed at "regularizing" Uber drivers certainly constitute a major incentive for platforms to negotiate collective agreements at national level. For instance, the Spanish collective agreement between Just Eat, CCOO and UGT (Table 6.1) was reached following the Spanish Supreme Court's decision on the employment status of workers in delivery companies and the so-called "Rider law" (as in the cases of similar agreements in Denmark and Italy). The

Table 6.1   Examples of social dialogue and collective bargaining in the platform economy

	Country	Year
• Collective agreement Syndicom and Swissmessengerlogistics	Switzerland	2019
In Switzerland, the trade union Syndicom signed a collective agreement with the Swissmessengerlogistics (SML) courier employers' association in February 2019, setting minimum standards for around 600 riders aimed at addressing precarious working conditions provided by companies like UberEats.		
• Collective agreement 3F and Hilfr	Denmark	2018
In April 2018, the Danish trade union 3F and the platform for cleaning services Hilfr signed the first collective agreement on platform work in Denmark. The agreement entered into force in August 2018 and ran as a pilot for 12 months. The collective agreement introduced a new category of worker – the so-called Super Hilfrs – in parallel with the existing freelance workers. Super Hilfrs are workers who opt for the status of employee rather than freelancer after meeting the eligibility criteria and will thus be covered by the company collective agreement. After working 100 hours, a worker automatically becomes a Super Hilfr (unless he or she be objects). Super Hilfrs receive a minimum hourly wage of DKK 141.21 (€19) and accrue rights to pensions, holiday entitlements and sick pay. Freelance workers' hourly wage is DKK 130 (€17) and they also receive a so-called "welfare supplement" of DKK 20 (€3) per hour. Both freelance workers and Super Hilfrs can set their hourly wage higher than the minimum wage on their individual profile on the platform. They are also covered by an insurance via the private insurance company Tryg. Tryg offers insurance solutions to six Danish-owned labour platforms, which include coverage for liability and accidents.		

	Country	Year
• Collective agreement Italian logistics sector	Italy	2017
In December 2017, a collective agreement was concluded in the Italian logistics sector that for the first time included food delivery riders in its contractual qualifications. The agreement was signed by the unions Confetra, Anita, Conftrasporo, Can-Fita, Transport Confartigianato, Sna-Casartigiani, and by employer organizations such as Claai and Filt Cgil. The agreement covered working time, the requirement for notice and compensation for changes in working schedules, and in case of illness. Following this agreement, the union Cgil proposed to start negotiating the algorithms of food delivery platforms that manage task allocation and schedules.		
• Agreement between Swedish Transport Workers' Union and Platform *Bzzt*	Sweden	2017
An agreement between the platform Bzzt, which offers an Uber-like service with electric mopeds, and the Swedish Transport Workers' Union allowed Bzzt drivers to be covered by the Taxi Agreement, which gives the workers access to the same standards as traditional taxi drivers. Unlike many platform companies, the drivers in Bzzt are offered marginal part-time contracts.		
• Agreement to collaborate between UK Union – GMB and Platform *Hermes*	United Kingdom	2019
In February 2019, the British courier company Hermes negotiated a new agreement with the GMB union, offering drivers a guaranteed minimum wage and holiday pay in a deal to provide trade union recognition.		
• Agreement to collaborate between Dutch Union – FNV and Platform *Temper*	Netherlands	2018
In 2018 in the Netherlands, the platform Temper, which matches demand and supply for staff in hotels, restaurants and cafés, approached the hospitality division of the largest Dutch union FNV (Federation National Unions, FNV-Horeca). This division of FNV and the platform signed a "cooperation pact" as a pilot scheme that would last one year to provide (legally self-employed) Temper workers with training, pensions and insurance. The cooperation between Temper and FNV-Horeca was broadened later in 2018 after a positive experience in the first months, adding further elements such as the removal of a software fee that Temper workers had to pay and improved training offerings.		

	Country	Year
• Agreement to collaborate between UK Association – IPSE and Platform Uber	United Kingdom	2017
In 2017, Uber partnered with the UK association IPSE (Independent Professionals and the Self-Employed) to provide discounted illness and injury insurance for Uber drivers. Drivers can avail themselves of this for £2 (about €2.20) per week, instead of the "market rate" of about £8 (€8.80) per week, and are insured in case of illness and injury for up to £2,000 (€2,200) if they are unable to drive for two or more weeks.		
• Agreement to collaborate between Australian Unions – NSW and Platform *Airtasker*	Australia	2017
An agreement between Unions NSW and Airtasker specified several basic practices and protections for workers, including recommended rates of pay, injury insurance, safety and dispute resolution. Indicatively: concerning minimum rates of pay, as of March 2017, Airtasker no longer posts any recommended pay rates below the 2016–17 National Minimum Wage for casual workers of $22.13 per hour; with regard to dispute settlement, Unions NSW, Airtasker and the Fair Work Commission agreed to develop an appropriate dispute resolution system to be overseen by the Commission, which would also act as the ultimate arbitrator. This is an important step in acknowledging the dependent nature of workers on the platform and the importance of an independent and transparent arbitration system in the case of disputes.		
• Seattle Ordinance giving drivers right to collectively bargain	USA	2015
In Seattle, the Teamsters Union (drivers of App-based transportation companies, such as Uber and Lyft) joined with local unions to press for a city Ordinance promoting collective bargaining between platform workers (despite their independent contractor status) and the transportation network company for whom they work (*Drivers' Collective Bargaining Ordinance*). The Seattle Ordinance does not take a position on whether the drivers are independent contractors or employees. Rather, the stated goal of the Ordinance is to "level the bargaining power between for-hire drivers and the entities that control many aspects of their working conditions." The law has been challenged repeatedly by platform-based companies as well as the US Chamber of Commerce.		

	Country	Year
• Agreement to collaborate between Danish HK PRIVAT and Platform *Voocali*	Denmark	2018
Voocali (a tech start-up, operating with freelancers, that has built an interpreter platform that can handle both video remote interpreting and on-site interpretation) signed the HK Agreement for Salaried Employees and a special agreement that covers work performed via the platform by those who are not employees. The parties agreed that freelancers are not paid below the bottom quartile for salary. They also discussed how to set aside funds for freelancers' further education through HK Privat's skills fund for freelancers. They then worked on reaching a pension agreement, so that freelancers can choose to have Voocali pay pension contributions to their pension savings.		
• Assodelivery, UGL on food delivery platforms in Italy	Italy	2020
The employers' organization (Assodelivery), representing the main food delivery platforms in the country, signed the first national collective agreement on the gig economy with the Unione Generale del Lavoro in September 2020. The agreement set a minimum hourly salary comprising actual delivery times, but also special allowances that cover night shifts, work during national holidays and bad weather.		

	Country	Year
• The Just Eat, CCOO and UGT agreement on food delivery workers in Spain	Spain	2021
In December 2021, the delivery platform Just Eat and the CCOO and UGT trade union confederations signed the first collective agreement for food delivery workers in Spain. The agreement, in force since January 2022, regulates wages and working time and conditions. Among other things, it sets annual wages for full-time riders, benefits for night work or work during holidays, time for vacations, maximum working time, provision by the platform of worktools (such as vehicle and mobile), compensation for riders using their own worktools, data protection privacy and providing information to workers' representatives about the algorithm used to manage their work.		
• The UK Uber – GMB agreement	UK	2021
In May 2021, in the UK, Uber signed a collective bargaining agreement with the trade union GMB. Through the agreement Uber has formally recognized GMB, which will now be able to represent up to 70,000 Uber drivers across the UK. The agreement aims at negotiating various work-related protections for Uber drivers, including living wage guarantee and holiday pay, pensions, discretionary benefits, health, safety and wellbeing, account deactivations, representation, and periodicity of social dialogue.		

Sources: Authors elaborating on Mexi (2019) and Swiss Network for International Studies (2019).

agreement also follows the European Commission's platform work directive proposal, which is based on a presumption of employment for all platform workers, including food delivery couriers, across the European Union. All in all, the shift to a more socially responsible profile seems to be mediated by country-specific institutionalized norms regarding appropriate corporate behaviour (Campbell 2007).

However, such processes of social dialogue and coordination among stakeholders have so far been observed primarily in localized platform work at national level, but not cross-border web-based crowdwork. With the right top-down incentive (for example, "shadow of regulation") and engagement of the relevant actors, particularly platform businesses, crowdworkers and their representatives, as well as governments, we may witness over time a more effective and concerted approach also regarding platform-based global crowdwork at cross-border level.

A Memorandum of Understanding (MoU) signed between Uber and the International Transport Workers' Federation (ITF) – a Global Union Federation – in February 2022 is one illustration of this approach. The MoU commits parties to begin social dialogue on platform workers' conditions and benefits globally. The parties agreed to start a cross-border social dialogue on collective bargaining, trade union representation, health and safety, working conditions, social protection, and litigation management. While the employment status of Uber workers – the most controversial point – is not discussed, the MoU signals a clear commitment of both parties to open roundtable discussions and to collaborate on measures to benefit drivers and couriers. Importantly, it is the first time that a dialogue between a platform company and trade unions takes place at a global level.

Such a development is likely to be a crucial step towards promoting a common labour relations framework at cross-border level, pending the creation of a fully fledged global inter-state regulatory framework such as the one proposed by the Global Commission on the Future of Work – that is, an "international governance system for digital labour platforms" conceived as a coherent and coordinated system, with universal and common minimum standards, that would apply to all workers irrespective of their location and employment status (ILO 2019c).

4.   CONCLUSION AND WAY FORWARD

The growth of geographically dispersed, cross-border, platform work presents serious challenges that cannot be effectively addressed without considering the need of establishing an international regulatory frame-work. In spite of the income and jobs opportunities digital labour plat-forms generate, regulatory gaps are likely to worsen working conditions for crowdworkers over time. Against this background, calls for strong representation and social dialogue across borders are becoming more and more pertinent, replicating the emerging experience of social dialogue and agreements observed at national level.

Cross-border social dialogue processes, such as in the context of transnational company agreements or sectoral collective bargaining, have evidenced the efficacy of self-regulatory initiatives in terms of promoting core international labour standards and addressing issues pertaining to highly globalized industries (such as the maritime transport or the textile and apparel sectors) and their supply chains. Some play a determinant role for determining wages and working conditions down the value chain of entire multinational companies and sectors. Granting continuance to these achievements, and building on them, would involve efforts to firmly embed cross-border social dialogue also in the platform economy and more specifically, online remote crowdwork – that is, the type of gig work where national regulation alone is clearly not enough. Importantly, the flexibility voluntary cross-border social dialogue offers could fit the evolving nature of this industry, which is based on a constantly evolving technology.

The agenda of such cross-border social dialogue – in the form of nego-tiations, consultations or exchange of information between global social partners leading to transnational company agreements and global sectoral collective agreements – could focus on several priorities (Albrecht et al. 2020; Kuek et al. 2015) including: pay and working conditions, data protection and privacy, transparent and fair contracts, social protection such as paid sick leave and pensions, freedom of association, worker rep-resentation, and labour-management consultation. Of course, this list is not exhaustive. The willingness of Global Union Federations – by far the most organized trade unions operating at the cross-border level – to adapt to the changing circumstances and to actively seek engagement with platform workers and businesses will be a decisive factor for introducing and cementing cross-border social dialogue and forms of self-regulation in the long run.

Voluntary cross-border social dialogue may also be a first step towards an international governance framework for digital labour platforms, if the ILO constituents so decided. And even once such a standard is adopted, cross-border social dialogue might remain an element of regulation, reinforcing each other. For instance, the adoption and periodical renegotiation of the global sectoral collective agreement in the maritime transport since 2003, have taken place against the background of the institutional framework of the ILO serving to set seafarers' minimum wages and to define other terms and conditions of employment for the sector through ILO Conventions and Recommendations.[10]

In its 2019 centenary year, a fresh mandate was given to the ILO on enhancing the contribution of cross-border social dialogue to the promotion of decent work, in an increasingly interconnected world of work. Cross-border social dialogue in the crowdwork industry could prove a credible pathway towards better working conditions and fairer terms of cooperation between platforms and crowdworkers. Looking ahead, this could lead to more ambitious global social regulation, more responsive to the needs of societies and economies in the 21st century.

NOTES

1. Another type of platform-based work, called "location-based", refers to platform work that involves tasks carried out in person in specific and easily identifiable locations (for example, Uber-type).
2. See https://tradingeconomics.com/uber:us:market-capitalization (accessed on 3 October 2022).
3. See ILO, "Decent Work in Global Supply Chains" (Report IV, International Labour Conference 2016); ILO, Resolution concerning decent work in global supply chains adopted on 10 June 2016 para 23(c); ILO, Meeting of experts on cross-border social dialogue, Final Conclusions, Geneva 12–15 February 2019 conclusion no 8.
4. See Prassl and Risak's seminal work on the multilaterality of work for multiple employers arising "particularly in the context of primarily digital crowdwork, when platforms, customers, and workers operate in different countries." As they explain, "there are multiple entities, and multiple modes in which these entities can share the exercise of employer functions" (Prassl and Risak 2016, pp. 621, 651).
5. By now, the literature has documented multiple initiatives implemented in relation to mobilization, representation and collective bargaining in the platform economy. An exhaustive discussion is beyond the scope of this chapter. The examples presented herein are intended to highlight associational forms that are gaining prominence for representing platform economy workers. More information on similar cases and initiatives can be found in the Eurofound web repository, https://www.eurofound.europa.eu/da/data/platform-economy/initiatives#organisingplatforms (accessed on 3 October 2022); see also Prassl (2018), Vandaele (2018) and Aloisi (2019).
6. See https://www.eurofound.europa.eu/data/platform-economy/initiatives/transnational-federation-of-couriers (accessed on 3 October 2022).

7. See https://ec.europa.eu/commission/presscorner/detail/en/ip_22_1145 (accessed on 3 October 2022).
8. See https://ec.europa.eu/commission/presscorner/detail/en/ip_21_6605 (accessed on 3 October 2022).
9. See http://faircrowd.work/unions-for-crowdworkers/frankfurt-declaration/ (accessed on 3 October 2022).
10. In particular, the ILO Joint Maritime Commission periodically recommends the minimum wage for a seafarer under the Seafarers' Wages, Hours of Work and the Manning of Ships Recommendation, 1996 (No. 187) (now consolidated within the Maritime Labour Convention, 2006). This recommendation, which is periodically updated based on negotiations within the Joint Maritime Commission, serves as an important benchmark for wage negotiations not only at national but also at international level – that is, in the International Bargaining Forum.

REFERENCES

Agrawal, A., J. Horton, N. Lacetera and E. Lyons (2013), "Digitization and the Contract Labor Market: A Research Agenda", NBER Working Paper 19525, NBER, Cambridge, MA.

Albrecht, T., K. Papadakis and M. Mexi (2020), "An International Governance System for Digital Labour Platforms", in *The Transformation of Work*, e-book published by Social Europe and Friedrich Ebert Stiftung. Available at: https://socialeurope.eu/book/the-transformation-of-work (accessed on 3 October 2022).

Aloisi, A. (2019), "Negotiating the Digital Transformation of Work: Non-standard Workers' Voice, Collective Rights and Mobilisation Practices in the Platform Economy", EUI MWP Working Paper 2019/03.

Baccaro, L. (2001), "Civil Society, NGOs, and Decent Work Policies: Sorting out the Issues", Decent Work Research Programme, DP/127/2001, Geneva: International Institute for Labour Studies (IILS)/ILO.

Berg, J. (2019), "Protecting Workers in the Digital Age: Technology, Outsourcing and the Growing Precariousness of Work", *Comparative Labour Law and Policy Journal*, **41** (1), 66–94.

Berg, J., A.M. Cherry and U. Rani (2019), "Digital Labour Platforms: A Need for International Regulation?", *Revista de Economía Laboral*, **16** (2), 104–128.

Berg, J., M. Furrer, E. Harmon, M. Rani and M.S. Silberman (2018), *Digital Labour Platforms and the Future of Work: Towards Decent Work in the Online World*, Geneva: International Labour Office.

Bergvall-Kåreborn, B., and D. Howcroft (2014), "Amazon Mechanical Turk and the Commodification of Labour", *New Technology, Work and Employment*, **29** (3), 213–223.

Bonvin, J.-M. (1998), *L'organisation Internationale du Travail: Etude sur une Agence Productrice de Normes*, Paris: PUF.

Bright, C. (2021), "Mapping Human Rights Due Diligence Regulations and Evaluating Their Contribution in Upholding Labour Standards in Global Supply Chains", in G. Delautre, E. Echeverría Manrique and C. Fenwick (eds),

Decent Work in a Globalized Economy: Lessons from Public and Private Initiatives, Geneva: ILO, pp. 75–108.

Campbell, J.L. (2007), "Why Would Corporations Behave in Socially Responsible Ways? An Institutional Theory of Corporate Social Responsibility", *Academy of Management Review*, **32** (3), 946–967.

Codagnone, C., F. Abadie, F. and F. Biagi (2016), "The Future of Work in the 'Sharing Economy': Market Efficiency and Equitable Opportunities or Unfair Precarisation?", Institute for Prospective Technological Studies, JRC Science for Policy Report EUR 27913.

da Costa, I., and U. Rehfeldt (2008), "Transnational Collective Bargaining at Company Level: Historical Developments", in K. Papadakis (ed.), *Cross-Border Social Dialogue and Agreements: An Emerging Global Industrial Relations Framework?*, Geneva: International Institute for Labour Studies (IILS)/ILO, pp. 43–64.

Davies, S., W. Glynne and N. Hammer (2011), "Organizing Networks and Alliances: International Unionism between the Local and the Global", in K. Papadakis (ed.), *Shaping Global Industrial Relations: The Impact of International Framework Agreements*, Geneva and New York: ILO & Palgrave Macmillan, pp. 201–219.

De Munck, J., C. Didry, I. Ferreras and A. Jobert (eds) (2012), *Renewing Democratic Deliberation in Europe: The Challenge of Social and Civil Dialogue*, Brussels: P.I.E. Peter Lang.

Degryse, C. (2016), "Digitalisation of the Economy and its Impact on Labour Markets", ETUI Working Paper 2016, Brussels: ETUI.

Didry, C., and A. Jobert (2011), "Social Dialogue and Deliberation: A New Dimension in European Industrial Relations", in J. De Munck, C. Didry, J.-I. Ferreras and A. Jobert (eds), *Civil Dialogue, Social Dialogue: A New Connection to Change our Model of Development*, Brussels: Peter Lang, pp. 173–89.

Drahokoupil, J., and B. Fabo (2016), "The Platform Economy and the Disruption of the Employment Relationship", ETUI Policy Brief No 5/2016, Brussels: ETUI.

Drahokoupil, J., and B. Fabo (2017), "Outsourcing, Offshoring and the Deconstruction of Employment: New and Old Challenges", in A. Serrano-Pascual and M. Jepsen (eds), *The Deconstruction of Employment as a Political Question*, Cham: Palgrave Macmillan, pp. 33–61.

Durward, D., I. Blohm and J.M. Leimeister (2016), "Crowd Work", *Business and Information Systems Engineering (BISE)*, **58** (4), 1–6.

Eurofound (2015), *New Forms of Employment*, Dublin: Eurofound.

Flecker, J., T. Riesenecker-Caba and A. Schönauer (2017), "Arbeit 4.0", Sozialbericht 2015–2016, Sozialministerium, Vienna, pp. 379–396.

Forlivesi, M. (2018), "Alla Ricerca di Tutele Collettive per i Lavoratori Digitali: Organizzazione, Rappresentanza, Contrattazione", *Labour and Law Issues*, **4** (1), 36–58.

Forsyth, A. (2019), "'Prova di Solidarietà': How Effectively are Unions and Emerging Collective Worker Representatives Responding to New Business

Models in Australia and Italy?", Paper for the 17th International Conference in Commemoration of Prof. Marco Biagi, Modena, 18–20 March.

Garben, S. (2019), "Tackling Precarity in the Platform Economy—and Beyond", *Social Europe*. Available at: https://socialeurope.eu/tackling-precarity-in-the -platform-economy-and-beyond (accessed on 3 October 2022).

Graham, M., and M.A. Anwar (2019), "The Global Gig Economy: Towards a Planetary Labour Market?", *First Monday*, **24** (4) [online paper].

Graham, M., I. Hjorth and V. Lehdonvirta (2017), "Digital Labour and Development: Impacts of Global Digital Labour Platforms and the Gig Economy on Worker Livelihoods", *SAGE Journals*, **23** (2), 135–162.

Heeks, R. (2016), "Examining 'Digital Development': The Shape of Things to Come?", GDI Development Informatics Working Paper No. 64, University of Manchester, UK.

Heeks, R. (2017), "Digital Economy and Digital Labour Terminology: Making Sense of the 'Gig Economy', 'Online Labour', 'Crowd Work', 'Microwork', 'Platform Labour', Etc." Development Informatics Paper 70, University of Manchester, UK.

Horton, J. (2010), "Online Labor Markets", in A. Saberi (ed.), *Internet and Network Economics*, Berlin: Springer, pp. 515–22.

Huws, U., N.H. Spencer and S. Joyce (2016), "Crowd Work in Europe: Preliminary Results from a Survey in the UK, Sweden, Germany, Austria and the Netherlands", FEPS Studies, Brussels: FEPS.

International Labour Office (2013a), *Social Dialogue Interventions: What Works and Why? A Synthesis Review 2002–2012*, Geneva: ILO.

International Labour Office (2013b), "Social Dialogue: Recurrent Discussion under the ILO Declaration on Social Justice for a Fair Globalization", Report VI, International Labour Conference 102nd Session, Geneva: ILO.

International Labour Office (2018a), "International Framework Agreements in the Food, Retail, Garment and Chemicals Sectors: Lessons Learned from Three Case Studies", Geneva: ILO. https://www.ilo.org/wcmsp5/groups/ public/---ed_dialogue/---sector/documents/publication/wcms_631043.pdf (accessed on 3 October 2022).

International Labour Office (2018b), "Social Dialogue and Tripartism: A Recurrent Discussion on the Strategic Objective of Social Dialogue and Tripartism, under the Follow-up to the ILO Declaration on Social Justice for a Fair Globalization", Report VI, ILC.107/VI, Geneva: ILO. https://www.ilo.org/wcmsp5/groups/public/---ed_norm/---relconf/documents/ meetingdocument/wcms_624015.pdf (accessed on 3 October 2022).

International Labour Office (2019a), "Cross-border Social Dialogue: Report for Discussion at the Meeting of Experts on Cross-border Social Dialogue" (Geneva, 12–15 February), MECBSD/2019, Geneva: IKI. https://www .ilo.org/wcmsp5/groups/public/---ed_dialogue/---dialogue/documents/ meetingdocument/wcms_663780.pdf (accessed on 3 October 2022).

International Labour Office (2019b), "Meeting of Experts on Cross-border Social Dialogue. Conclusions" (Geneva, 12–15 February), Geneva: ILO. https://www.ilo.org/wcmsp5/groups/public/---ed_dialogue/---dialogue/

documents/meetingdocument/wcms_700599.pdf (accessed on 3 October 2022).

International Labour Office (2019c), *Work for a Brighter Future: Global Commission on the Future of Work*, Geneva: ILO.

International Labour Office (2021), *World Employment and Social Outlook: The Role of Digital Labour Platforms in Transforming the World of Work*, Geneva: ILO.

Keck, M. E., and K. Sikkink (1998), *Activists beyond Borders: Advocacy Networks in International Politics*, Ithaca, NY: Cornell University Press.

Kenney, M., and J. Zysman (2015), "Choosing a Future in the Platform Economy: The Implications and Consequences of Digital Platforms", Kauffman Foundation New Entrepreneurial Growth Conference, Discussion Paper, Amelia Island Florida, 18 June.

Kuek, S. C., C. Paradi-Guilford, T. Fayomi, S. Imaizumi, P. Ipeirotis, P. Pina and M. Singh (2015), *The Global Opportunity in Online Outsourcing*, Washington DC: World Bank.

Lehdonvirta, V., H. Barnard, M. Graham and I. Hjorth (2014), "Online Labour Markets: Levelling the Playing Field for International Service Markets?", Oxford Internet Institute, Oxford, UK.

Leimeister, J.M., D. Durward and S. Zogaj (2016), "Crowd Worker in Deutschland: Eine empirische Studie zum Arbeitsumfeld auf externen Crowdsourcing-Plattformen". Available at: https://www.researchgate.net/publication/307546836_CROWD_WORKER_IN_DEUTSCHLAND_Eine_empirische_Studie_zum_Arbeitsumfeld_auf_externen_Crowdsourcing-Plattformen (accessed on 3 October 2022).

Margaryan, A. (2016), "Understanding Crowdworkers' Learning Practices", paper presented at Internet, Politics, and Policy 2016 conference, Oxford, 22–23 September.

Mexi, M. (2019), 'Social Dialogue and the Governance of the Digital Platform Economy: Understanding Challenges, Shaping Opportunities', background paper for discussion at the ILO-AICESIS-CES Romania International Conference (Bucharest, 10–11 October 2019). Available at: https://www.ilo.org/wcmsp5/groups/public/---ed_dialogue/---dialogue/documents/meetingdocument/wcms_723431.pdf (accessed on 3 October 2022).

Mexi, M. (2020), 'The Platform Economy—Time for Decent Digiwork', Social Europe, November 2020. Available at: https://socialeurope.eu/the-platform-economy-time-for-decent-digiwork (accessed on 3 October 2022).

OLI (2020), Online Labour Observatory, onlinelabourobservatory.org (accessed on 3 October 2022).

Organisation for Economic Co-operation and Development (2019), *Employment Outlook 2019: The Future of Work*, Paris: OECD.

Papadakis, K. (2006), *Civil Society, Participatory Governance and Decent Work Objectives: The Case of South Africa*, Geneva: ILO/IILS.

Papadakis, K. (2011a), "Globalizing Industrial Relations: What Role for International Framework Agreements?", in S. Hayter (ed.), *The Role of Collective Bargaining in the Global Economy: Negotiating for Social Justice*,

Geneva, and Cheltenham, UK and Northampton, MA, USA: ILO and Edward Elgar, pp. 277–304.

Papadakis, K. (ed.) (2011b), *Shaping Global Industrial Relations: The Impact of International Framework Agreements*, Geneva and New York: ILO and Palgrave Macmillan.

Papadakis, K. (2021), "A Short History and Future Prospects of Cross-border Social Dialogue and Global Industrial Relations Agreements", in G. Delautre, E. Echeverria and C. Fenwick (eds), *Decent Work in a Globalised Economy: Lessons from Public and Private Initiatives*, Geneva: ILO, pp. 133–161.

Papadakis, K., G. Casale and K. Tsotroudi (2008), "International Framework Agreements as Elements of a Cross-border Industrial Relations Framework", in K. Papadakis (ed.), *Cross-border Social Dialogue and Agreements: An Emerging Global Industrial Relations Framework?*, Geneva: International Institute for Labour Studies (IILS)/ILO, pp. 67–87.

Prassl, J. (2018), *Collective Voice in the Platform Economy: Challenges, Opportunities, Solutions*, Brussels: ETUC.

Prassl, J., and M. Risak (2016), "Uber, Taskrabbit, & Co: Platforms as Employers? Rethinking the Legal Analysis of Crowdwork", *Comparative Labor Law and Policy Journal*, **37** (3), 619–651.

Rehfeldt, U. (1998), "European Works Councils: An Assessment of French Initiatives", in W. Lecher and H.-W. Platzer (eds), *European Union: European Industrial Relations? Global Challenges, National Developments and Transnational Dynamics*, London and New York: Routledge, pp. 207–222.

Salehi, N., L.C. Irani, M.S. Bernstein, A. Alkhatib, F. Ogbe, K. Milland and Clickhappier (2015), "We Are Dynamo: Overcoming Stalling and Friction in Collective Action for Crowd Workers", Proceedings of the 33rd Annual ACM Conference on Human Factors in Computing Systems, Seoul, Republic of Korea, 18–23 April.

Schmidt, F.A. (2017), *Digital Labour Markets in the Platform Economy*, Bonn: Friedrich Ebert Stiftung.

Stephany, F., O. Kässi, U. Rani and V. Lehdonvirta (2021), "Online Labour Index 2020: New Ways to Measure the World's Remote Freelancing Market", *Big Data & Society*. Available at: https://journals.sagepub.com/doi/10.1177/20539517211043240 (accessed on 3 October 2022).

Swiss Network for International Studies (2019), Final Working Paper on *Gig Economy and Its Implications for Social Dialogue and Workers' Protection*. Available at: https://snis.ch/wp-content/uploads/2020/01/Working-Paper-Gig-Economy.pdf (accessed on 3 October 2022).

Vandaele, K. (2018), "Will Trade Unions Survive in the Platform Economy? Emerging Patterns of Platform Workers' Collective Voice and Representation in Europe", ETUI Research Paper, Working Paper 2018.05.

7. Conclusion: The rise and growth of the gig economy. Challenges and opportunities for social dialogue and decent work

Jean-Michel Bonvin, Nicola Cianferoni and Maria Mexi

History has shown the importance of workers' collective organization for the enhancement of working conditions and the setting up of adequate labour and social protection mechanisms (Supiot 2012). The International Labour Organization (ILO) has emphasized this dimension, by insisting that trade union freedom and workers' ability to organize and to collectively bargain working conditions and social protection devices are prerequisites for the implementation of decent work defined as work "that is productive; ensures equality of opportunity and treatment for all women and men; delivers a fair income, security in the workplace and social protection for families; provides prospects for personal development; and gives workers the freedom to express their concerns, organize and participate in decisions that affect their working lives" (Berg et al. 2018, p. 1). In this perspective, the ILO Decent Work Agenda has been framed around four pillars: fundamental rights at work as stated in the 1998 Declaration on Fundamental Principles and Rights at Work (trade union freedom and right to organize and collective bargaining as developed in ILO Conventions 87 and 98 are on top of the list of fundamental rights), social dialogue, job creation and social protection. In the global and virtual world of work, such emphasis on social dialogue and workers' collective organization is even more relevant if the challenge of decent work is to be met. Our book investigates the opportunities and obstacles faced by social dialogue and collective organization in the context of the gig economy. In this chapter, we synthesize main findings around two main issues. First, is there a shared diagnosis about the issue at stake? More specifically, is the gig economy perceived as a problem calling for innovative solutions? Or can it be tackled via minor adjustments of existing regulations? Or, yet another perception, should it rather be conceived as an opportunity to ask for more flexible regulations? Such diversity of viewpoints significantly impacts the perceived possibilities of social dialogue to achieve decent digi-work (Mexi 2020) and the strategies implemented to this purpose. Second, how does the variety of actors (gig workers, trade unions, employers' associations, public authorities) involved at various levels (local, sectoral, national and international) connect to each other? Taking stock of the previous chapters, we show how public and private actors interact at each relevant level and how they are able to overcome (or reinforce, if such is their strategy) the obstacles represented by the diverse forms of fragmentation identified in the introduction (Heiland 2020). A contrast can indeed be observed between strategies aiming at reinforcing workers' voices in the collective bargaining process and strategies where the objective is to silence workers (Kougiannou and Mendonça 2021). The following issues are more specifically tackled: Do gig workers succeed in building collective organizations and, if so, how do these relate to existing trade unions (Joyce and Stuart 2021)? How do conventional employers relate to employers in the gig economy: do they perceive them as unfair competitors or as potential allies in collective bargaining processes? How do social partner mobilizations impact political discussions and debates at national level? How do all these grass-roots or more institutional initiatives at national level translate into actions at international or global level?

1.   A SHARED DIAGNOSIS, BUT MULTIPLE INTERPRETATIONS OF ITS IMPLICATIONS

In all countries investigated in this book, there is an overall shared diagnosis about the precarious working conditions entailed by the gig economy, as well as its implications in terms of access to adequate labour and social protection. This applies to all sectors investigated (accommodation, cleaning, delivery and transportation services), as well as to the two main forms of gig work: on-location gig work where the actual work is achieved in the real world although the intermediation takes place online, and online crowdwork where work can be accomplished online, in the virtual world (see also Berg et al. 2018). This confirms the assumption that the platform economy introduces "a new organization form, stepping in as an intermediary in increasingly broader kinds of work, collecting both data and a cut of the payments made for services" (Woodcock & Graham 2020, p. 61), which contributes to degraded working conditions. This has been exacerbated by the fact that most gig workers are not considered as dependent workers and thus do not benefit from rights and entitlements associated with labour. More specifically they are excluded both from social protection (as access to benefits is to a large extent tied to the worker status) and from social dialogue (as collective organization of self-employed workers is interpreted as a breach of competition law), that is from two pillars of decent work. Moreover, though claiming to respect the workers' autonomy in the daily work organizations, the platform controls the transaction that it organizes and introduces a significant asymmetry of power, notably via the tools of algorithmic management (Wood et al. 2019). Thus, the gig or platform economy represents today a very big challenge for the implementation of the ILO Decent Work Agenda, as the ILO Global Commission on the Future of Work has confirmed (ILO 2019).

However, the implications of this diagnosis vary significantly along the sectors and countries investigated. While in the transportation sector the emergence of the gig economy is primarily envisaged as a labour issue, in the accommodation sector it is framed as a matter of (un-)fair competition. This impacts the ways collective organization is implemented: in the transportation sector, this is mostly an issue of mobilizing gig workers – either through grass-roots initiatives or via incentives from existing trade unions – and creating an efficient coalition between gig and traditional workers; in the accommodation sector, the mobilization rather concerns traditional employers and platform economy managers and mostly focuses on fiscal matters. Thus the parties and positions involved vary according to the salience of the main – labour or competition – issue. Working conditions and precariousness are considered as the most pressing problem in the transportation and delivery sectors insofar that these activities are considered under the lens of labour. By contrast, being active in the accommodation sector is not perceived as work, but as one way to get a supplementary income: issues of labour and social protection are not perceived as relevant in this context, hence the problematic issue is that of unfair competition with the professionals of the accommodation sector. Also worth mentioning is the fact that not all gig workers perceive their precariousness as a problem, some categories seeing it as a way to complement other incomes (in certain situations it can even improve their situations as illustrated by the cleaners in Switzerland) or as transitional situations that are assessed as acceptable due to their short-term nature (typically students). Thus, gig workers' precariousness is not equally perceived as a problem by all workers and in all sectors. This may be a factor of social fragmentation when it comes to creating the conditions for collective action and mobilization. Such a variety of interpretive patterns can be observed throughout all the countries investigated.

However, countries differ along their system of industrial relations and these differences contribute to explain why similar problems (especially contract precariousness, low wages, limited access to social benefits) are perceived in a different way along diverse national industrial relations configurations. In countries with a strong focus on social dialogue and with social protection mechanisms that are quite well developed while allowing for flexibility on the labour market, those who want to preserve the social protection components of the system oppose those who prefer to maintain or increase its flexibility (or to pursue a flexibilizing trend that has already been initiated). This is mostly the case with Germany and Switzerland, where the compromise solution between both parties consists in a tendency towards the preservation of existing equilibria and, thus, a shared conviction that the gig economy need not result in a disruption of the existing model, though for different reasons: while trade unions and parties on the left want to preserve the protective components of the model, suggesting, for example, the implementation of existing labour law provisions to gig and traditional workers alike, employers and parties on the right insist on preserving the flexibility of the existing pattern of industrial relations. This results in incremental change and a high degree of stickiness of established institutions of social partnership. Thus, digitalization and the gig economy are interpreted as yet another instance of the classic conflict between labour and capital: while there seems to be an insurmountable disagreement about gig workers' status and the necessity to further (de-)regulate the labour market, this does not question the existing compromise between social partners.

In Germany and Switzerland, there is overall support for the development of digitalization and the platform economy (emphasizing the need for flexibility so that job creation opportunities can be taken, or the necessity to adapt training curricula to digital requirements), together with a wide acceptance that issues related to gig workers' precariousness can be solved within the present legal framework rather than create a third hybrid status or modify the balance of rights and obligations attached to the existing worker status. This does in no way mean that the present framework is considered an ideal solution by all partners, but it is an acceptable compromise in their eyes, with therefore a higher likelihood of being accepted than more disruptive solutions supported only by one of the parties involved. Two dividing lines can be observed: parties on the left and trade unions vs. employers' associations and parties on the right on the one hand; traditional employers and workers vs. gig employers and workers on the other hand. In such contexts, the risk is that the preference for consensual solutions may result in reinforcing insider–outsider dilemmas, where the protection of traditional workers prevails over that of gig workers. No wonder then if decision-makers in political arenas do not succeed in reaching compromise solutions with regard to the key issue of the worker status, with the implication that such issues are then ruled out in court litigations examining on a case-by-case basis whether the degree of subordination observed in a specific employment relationship qualifies it as an employer–employee relationship or justifies that a self-employed status is granted. The implications are very significant: it determines whether the entrepreneurial risk is borne by the platform as an employer or by the worker him/herself identified as self-employed; but they do not coincide with a disruption of the existing system of industrial relations.

In the other two countries investigated, the status quo is not perceived as an acceptable compromise by all parties involved, though for different reasons. The higher levels of income inequality and the extent of employment insecurity that can be observed in the overall UK labour market design a specific environment, where the trends of precariousness and casualization characteristic of the gig economy reinforce preexisting tendencies that are also favoured by a widely pro-market model of industrial relations with an employer-oriented pattern of social partnership. In such a context, trade unions are less supportive of the existing model and tend to use the gig economy as an emblematic case in order to contest some aspects of this model. This may result in openly conflictual industrial relationships and public protests, with a view to changing the law rather than including gig workers in the scope of the extant labour law provisions. Actions taken in the workplace or via the courts, mostly deriving from grass-roots initiatives, are meant to advance the protection of workers. Gig workers themselves have a very negative perception of their working conditions, which may translate into radical forms of action, while established trade unions are torn between the will to enhance overall workers' protection and the ambition to keep their privileged contact with institutional partners. In Greece, even though the status quo is not perceived as attractive, there is no real attempt at challenging it. The model of industrial relations is perceived as unsatisfactory, mostly due to austerity policies following the 2008 crisis and historical legacies of clientelism and political patronage resulting in fragmented interest structures. In such circumstances, social partners' ability to address the austerity challenge and its deleterious implications for workers appears highly questionable. If the problems of precariousness and insecurity characteristic of the gig economy are identified, trade union action is not considered as a solution by the actors concerned, as is illustrated by the low scope of collective action and social dialogue in the Greek gig economy. Faced with a failing state (mostly constrained by the requirements of its creditors at the time of our empirical work) and to a large extent inactive trade unions, despite the poor quality of working conditions, the gig economy is envisaged as an attractive solution to provide extra income: since the welfare state is not protective enough and since social partners seem unable to address the negative features of the platform economy, it is paradoxically the gig economy itself that appears as a solution to improve, even though to a limited extent, the daily lot of Greek people. In this case, status quo is not endorsed as an acceptable compromise, but as a situation to which one somehow resigns oneself, using the gig economy to generate extra income. The potential for contesting the existing model of industrial relations and its limitations seems very limited under such circumstances.

Online crowdworkers, completing micro-tasks for digital platforms, find themselves in a similar situation: they accept very low wages that provide them with extra money without claiming to change the business model of such platforms. Both cases – location-specific gig workers in Greece and crowdworkers broadly – show that an objective diagnosis of poor working conditions need not translate into collective action aimed at contesting such conditions; it may well be the case that people resign themselves to these situations, especially when they are faced with the inability of the state to protect them and of social partners to address their problems. In the case of crowdwork, this inability is reinforced by the high risks of spatial, organizational and technological fragmentation, as we will develop in the next section.

2. COLLECTIVE ACTION AND SOCIAL DIALOGUE FACED WITH THE RISKS OF FRAGMENTATION AND THE DIFFICULTIES OF MULTILEVEL COORDINATION

When investigating the translation of diagnoses and their interpretations into collective action and mobilization, three levels or types of actors must be taken into account, which may act in a complementary (though full convergence in the same direction is exceptional and has not been observed in our research) or in a compensatory way. The latter case is by far the most frequent, as action at one level may compensate (or rather try to compensate) inaction or apathy at one or more of the other levels.

The first level is that of the gig platforms. At this level, we can observe silencing strategies, on the one hand, and voicing strategies, on the other hand. Silencing strategies aim at preserving the business model of platforms from the protesting voices of workers. In the investigated case studies, this is achieved in two main ways. First, there are attempts at framing the gig economy as a "non-problem" in terms of labour or as a situation where opportunities outweigh risks and should therefore not be impeded by excessive regulation. The accommodation sector represents a successful instance of framing the gig economy as a non-labour issue. Interestingly, protesting voices do not come from workers in this case, but from competitors in the sector who report unfair competition conditions and ask that platforms be submitted to the same professional requirements, tax conditions and the like. In the transportation sector, silencing strategies mostly rely on the imposition of the self-employed status of gig workers, thus framing the employment relationship in terms of contract law rather than labour law (which also excludes the right to collective organization as it is interpreted as a breach of competition law). In such cases, contractors do not have to abide by labour law provisions and can decide more freely about the content of labour contracts. Such silencing strategies are opposed both by conventional employers who have to comply with legal labour standards and thus incur a competitive disadvantage with regard to gig employers, and by gig workers themselves who contend that they are employees and not self-employed. This situation can be found in all countries investigated in this book, though with different outcomes in terms of workers' access to social protection and social dialogue. In short, "Move along, there is nothing to see here" is the message conveyed by this first category of silencing strategies.

The second category concerns situations where gig workers are recognized as wage earners and where labour standards apply: in this case, the objective is to promote private arrangements within the firm as much as possible. In other words, it is recognized that the gig economy includes labour issues, but these should be tackled at firm level and the state or trade unions should not interfere. Such strategies can be found specifically in the notime case in Switzerland as well as in the UK. Facing the failure to externalize labour costs, managers accept to internalize them provided this can be achieved on their own terms. The strategy is to silence external stakeholders so that labour management is envisaged as an internal task and can be based on private arrangements, where workers' voices can be more easily checked by managers. This second category of silencing strategies is increasingly important in the transportation and delivery sectors, where the employee status is more and more often granted by court decisions settling this issue on a case-by-case basis. This shows that the story does not end with the granting of an employee status, and that the implementation of decent digi-work remains a challenge even when gig workers are granted the wage earner status and all rights and obligations attached to it.

The success of silencing strategies depends to a large extent on the power of opposing strategies that aim at "breaking the managerial silencing of worker voice in platform capitalism", in the words of the title of Kougiannou and Mendonça's article (2021). Such voicing strategies, as we suggest naming them, require both willingness and capacity in order to set them up and implement them. Evidence from previous chapters suggests that willingness very much depends on how the gig economy contribution to workers' material well-being is perceived. If it represents an improvement vis-à-vis a previous situation that was even more precarious, the willingness to contest working conditions in the gig economy is reduced, as is illustrated by the cases of Greece and the Swiss cleaners. Such hesitation to contest the gig economy and claim for an employee status or other improvements of working conditions is further reinforced by the platform managers' threat to quit the country, which would leave those workers without that source of extra income. By contrast, when the gig economy is considered an infringement on decent work and a factor of increasing precariousness, workers are more willing to take the risks associated with collective organization and mobilization and disregard the platforms' threats. This can be observed more frequently in the delivery and transportation sectors.

Workers' capacity to collectively organize and, where necessary, override the fact that self-employed persons are not entitled to collective bargaining and social dialogue, depends on their ability to overcome the risks of fragmentation identified in the introduction of this book. In location-based gig work, evidence collected in this volume indicates that those obstacles are far from unsurmountable. Multiple strategies are used to overcome spatial, organizational and technological fragmentation: due to their visibility, delivery riders have for instance the possibility of meeting in the public space, thus overcoming the absence of formal meeting places and communication channels provided by the platform; workers also develop their own communication channels via the creation of WhatsApp groups or other autonomously operated chat groups, thus demonstrating their ability to use technology for their own purpose; they also resort to other more conventional tools such as public campaigning, petitioning, media releases, to make their voices heard, etc. These demonstrate gig workers' capacity to overcome these forms of fragmentation and find the means to create collective organization and solidarity rather than competing for gigs as the business model of platforms would have it. Social fragmentation cannot be considered an insurmountable obstacle either: in certain cases, workers who take some advantage of the gig economy, due to working time flexibility or to extra income, do not hesitate to take sides with their colleagues, thus showing that people do not behave only on the grounds of self-interest. Case studies in the Swiss and UK chapters illustrate that social fragmentation is not a fatality in the gig economy and that a sense of collective solidarity, beyond self-interest, can flourish among gig workers, in the same way as in all other economic sectors. In recent years, workers' capacity to overcome fragmentation and create collectives in the context of the location-based gig economy has been abundantly demonstrated (Tassinari and Maccarrone 2020; Dufresne and Leterme 2021). The increasing number of legal cases where gig workers' self-employed status is contested (often successfully) also shows that the legal fragmentation resulting from this status can be overcome.

By contrast, online crowdworkers have more difficulties in overcoming the risks of fragmentation, mainly due to the lack of interpersonal knowledge that makes it more difficult to implement collective solidarity and the exacerbation of legal fragmentation represented by the possibility for platforms to locate their activities in the digital space and thus implement a form of "regime shopping". In other words, platforms have a more readily available exit option that reinforces the prevalence of legal fragmentation in the case of crowdwork.

The second level relates to how gig workers' grass-roots mobilizations connect to actions taken by established social partners – that is, trade unions and employers' associations. When it comes to creating the conditions for worker collective action, the challenge is that of establishing complementarity between grass-roots initiatives and traditional trade unions action. As a matter of fact, the connection between grass-roots gig workers' movements and traditional trade unions is far from straightforward: for instance, Joyce and Stuart (2021, p. 179) indicate that only

34 percent of platform worker protests in Europe between January 2017 and June 2020 involved mainstream unions. This shows potential discrepancies between grass-roots unions' agenda, often politically more radical and endorsing a more conflictual mode of industrial relations, and long-standing unions, which tend to prefer negotiated solutions and are reluctant to put their institutional position at risk. Joining forces requires then to accommodate this variety of objectives. The list of collective agreements negotiated by traditional unions for gig workers that can be found in Chapter 6 (Table 6.1, pp. 131–135) illustrates the potential for such coalitions and how they can result in extensions of worker rights to social protection and social dialogue for part of the platform workforce and under certain conditions. The adoption of such solutions relies on platforms' willingness to enter such agreements, for instance in order to attract skilled workers via offering them more attractive conditions or to improve the firm's reputation vis-à-vis socially-sensitive customers. For this reason, these agreements often represent partial improvements and risk creating a two-tier workforce where the most skilled and highest performing enjoy greater entitlements and rights. They, however, represent a clear step forward from the strategy initially pursued by many established trade unions that were tempted to focus on their constituency and pay reduced attention to precarious workers' situations; with these emerging initiatives, trade unions become more active in the gig economy but with an emphasis on those working conditions and categories of gig workers for whom a compromise can be more easily found with gig employers. While representing a step forward, such efforts may come with some tensions with grass-roots movements, as is illustrated by the difficulties to create long-term complementarity between gig workers' mobilizations and trade union actions in the UK and in the notime case in Switzerland.

Social dialogue is only one component of the trade union strategy with regard to gig work: trade unions are indeed aware of the limitations of collective bargaining, mostly due to its reliance on employers' willingness to take part in these processes, and they complement it by resorting in the same time to court litigation and policy pressure strategies. Actually, the worker status issue is mostly debated in court litigations on a case-by-case basis. As illustrated in the previous chapters, the combination of these three strategies – social dialogue, court litigation and policy pressure – is implemented in Germany, the UK and Switzerland (especially in the delivery and transportation sectors), where it contributes to reinforce complementarity between grass-roots and mainstream trade unions, although there is still some way to go.

With regard to employers' associations, the observed pattern is more of a compensatory nature, meaning that gig workers' inaction may sometimes be compensated by action taken by traditional employers. This is the case in the accommodation and the cleaning sectors, where employers' action to redress unfair competition practices may result in some degree of enhancement of working conditions. Pressured by their competitors, platform managers are called to take measures that comply with the rules of fair competition and sometimes result in enhancing working conditions. This is also illustrated by the initial stage of the Uber case in canton Geneva, where a discussion between traditional employers (taxi operators) in the field and legislators initiated a movement toward an enhanced social protection for Uber drivers, which was joined at a later stage by established trade unions and emerging grass-roots movements. This shows that the field of the gig economy is riddled with a diversity of tensions both among workers (mostly between gig and traditional workers, along the insider–outsider dilemma line) and among managers (with a potential conflict between traditional and gig employers, along the unfair competition issue). In this context, a variety of coalitions can emerge, which may go in the direction of enhancing social protection or, on the contrary, of reinforcing employment flexibility. Cases investigated in this book illustrate the variety of potential outcomes that can result from this complex configuration.

The third category concerns policy actors at national and international level and the capacity of social partners to convince them to address issues of social protection and social dialogue in the gig economy. As was already mentioned earlier, the impact of collective mobilizations at national level does not translate into reinforcing existing labour market regulations, but rather contributes to preserve them from further deregulation processes. Hence, at national level, the improvement of gig workers' working conditions and their enhanced access to social rights and social dialogue are not achieved via policy reforms but through court litigations: this implies that such improvement takes place only for those whose subordinated employment status is confirmed by judicial decisions. All others will remain subject to their existing precarious working conditions.

Advocacy for regulatory reform through lobbying or campaigning also takes place at international level. That is the objective of the many attempts at building transnational collective action (Dufresne and

Leterme 2021) that are particularly successful in the delivery sector. Such efforts have certainly contributed to boosting the discussion at EU level, which resulted in the platform work directive proposal issued in the end of 2021. This document puts forward three central issues: first the necessity to address the risk of misclassification of gig workers as self-employed, by instituting a so-called employment presumption in all platforms exercising control over their employees while admitting the possibility that platforms may hire genuine self-employed workers; second, a call for more transparent algorithmic management via better information about monitoring and decision systems and their impact on working conditions; third, the setting up of a labour inspectorate in charge of enforcing the proposed directive and ensuring the transparency and traceability of platform work. It may be the case that this directive proposal will boost voluntary cross-border social dialogue initiatives by social partners who are willing to self-regulate rather than abide by such external regulations. At global level, there are also suggestions about potential ILO initiatives that could encourage cross-border social dialogue while providing a coherent international governance system ensuring that universal minimum standards apply to all workers, whatever their location or employment status. There are certainly promises of complementarity between such top-down (emanating from the EU or the ILO) and bottom-up (voluntary cross-border social dialogue, ideally including also gig workers' representatives) approaches, although there is still quite a long way to go.

3.   FINAL REMARKS

The gig economy has received increasing public attention over the past few years. But how can workers in the gig economy have their interests represented and bargain for better pay and working conditions? Our book has built on the pressing need to enhance context-sensitive knowledge on how the gig economy can become a catalyst for decent, fulfilling work in the modern labour market. By looking at recent developments in four European countries – Germany, Greece, Switzerland and the United Kingdom – it has examined how governments and the social partners perceive the impact of gig work on the labour market and how they shape responses in this regard. In a nutshell, the main findings point to:

(a)   An overall consensus about precarious working conditions in the gig economy, together with reduced access to social protection and social dialogue rights. From this widely shared diagnosis, a variety of implications are drawn according to national contexts and models of industrial relations, sectoral reality and gig workers' profiles. This also points to the irrelevance of one-size-fits-all solutions – in this sense solutions inspired by local or sectoral social dialogue seem more in line both with the needs of gig workers and with the competitiveness requirements of platform businesses. Under such circumstances, it is no wonder that the national case studies explored in this volume show a significant discrepancy between policy-makers and social partners' views about the issues raised by the gig economy (mostly focusing on the worker status or on issues related to tax avoidance or unfair competition), and the actual practices and experiences of gig workers themselves.

(b)   The relatively low use – with variation that is context- and sector-specific – of social dialogue within the gig economy field. Our case studies allowed identifying three main situations in this respect: first, a situation of absent social dialogue where gig work is not framed as a social problem, and workers have to devise their own solutions and individual arrangements to improve their working conditions; second, a situation of fragmented social dialogue where the mobilization of gig workers, when it takes place under such circumstances, stands in (sometimes sharp) contrast with existing trade unions defending their members' interests that they perceive as threatened by the platforms' business model; third, a situation of joint or unified social dialogue, where all workers, be they members of conventional trade unions or gig workers, transcend the boundaries of the insider–outsider dilemma to implement an inclusive view of collective bargaining. Also worth mentioning with regard to social dialogue in the gig economy is the emergence of new forms of mobilization such as the use of instant messaging groups or the gig strike that took place in the UK. These evolutions point to a renewal of social dialogue forms in order to adjust them to the specific circumstances of gig workers.

(c)   Evidence of low-to-moderate mobilization on the part of traditional social partners. This suggests that the first two forms of social dialogue mentioned in the previous point tend to prevail in the countries and sectors under investigation. This also points to the existence of a discrepancy between traditional employees' interests and gig workers' claims, as well as between existing trade unionists' competencies and those technological abilities that would be needed to tackle algorithmic and data-driven management. Such discrepancies in interests and competencies tend to reinforce insider-outsider dilemmas and their undesired implications. The promotion of decent digital work requires dissolving such boundaries. In recent years, an increasing number of initiatives pursue this objective.

(d) The lack of coherent full-fledged policy and legal responses at national level. This is above all an issue of framing: in all investigated countries and sectors, the gig economy is framed mainly as a problem of worker status, or tax avoidance and unfair competition. When investigating the gig economy from a grass-roots perspective, other issues such as algorithmic management, but also health and safety, minimum pay, etc. come to the fore. To close the gap and come to a more adequate and complete framing of the issues raised by the gig economy, in-depth scientific analyses are needed. This calls for a wide-ranging collaboration within interdisciplinary teams, including lawyers, social policy scholars, industrial relations scholars, occupational health experts, etc. Besides, our study has also shown that policy-makers' action, although important, will not be able to tackle all issues through the adoption of legislation; social dialogue is a necessary complement to public policy, as well as the court litigation strategy, and public policy needs to support such dialogue by providing adequate procedural rights – that is, the right to effectively take part in collective bargaining, to express one's voice and make it count – to all workers, including gig workers. As emphasized by the ILO, the promotion and implementation of decent work is all stakeholders' business and responsibility. Regulatory initiatives at international level in the line of the EU platform work directive proposal or following the model of the ILO Maritime Labour Convention, could helpfully complement the actions taken at national level. This shows not only the need, but also the possibility to strengthen cross-border social dialogue initiatives and agreements such as the transnational company agreements.

Looking ahead, the potential benefits of digitalization will accrue to all involved only once social dialogue is promoted and implemented as an invaluable mechanism for bringing the gig economy into the scope of national policy and regulatory interventions and adaptations, thus creating the preconditions for a fair and equitable market order in the gig economy and, more broadly, the whole economy (Deakin and Supiot 2009). This is the core conviction that inspired our book.

REFERENCES

Berg, J., Furrer, M., Harmon, E., Rani, U., & Silberman, M. S. (2018), *Digital Labour Platforms and the Future of Work: Towards Decent Work in the Online World*, Geneva: International Labour Office.

Deakin, S., & Supiot, A. (eds) (2009), *Capacitas, Contract Law and the Institutional Preconditions of a Market Economy*, Oxford and Portland: Hart Publishing.

Dufresne, A., & Leterme, C. (2021), *App Workers United: The Struggle for Rights in the Gig Economy*, The Left in the European Parliament, Brussels: European Parliament.

Heiland, H. (2020), "Workers' voice in platform labour", Study No. 21, Hans-Böckler-Stiftung, July.

International Labour Office (2019), *Work for a Brighter Future: Global Commission on the Future of Work*, Geneva: ILO.

Joyce, S., & Stuart, M. (2021), "Trade union responses to platform work: An evolving tension between mainstream and grassroots approaches", in J. Drahokoupil and K. Vandaele (eds), *A Modern Guide to Labour and the Platform Economy*, Cheltenham, UK and Northampton, MA, USA: Edward Elgar Publishing, pp. 177–192.

Kougiannou, N. K., & Mendonça, P. (2021), "Breaking the managerial silencing of worker voice in platform capitalism: The rise of a food courier network", *British Journal of Management*, 32 (3), 744–759. https://doi.org/10.1111/1467-8551.12505.

Mexi, M. (2020), "The platform economy—time for decent 'digiwork'", *Social Europe*, 26 November.

Stanford, J. (2017), "Historical and theoretical perspectives on the resurgence of gig work", *Economic and Labour Relations Review*, 3 (28), 328–401.

Supiot, A. (2012), *The Spirit of Philadelphia: Social Justice vs. the Total Market*, London: Verso.

Tassinari, A., & Maccarone, V. (2020), "Riders on the storm: Workplace solidarity among gig economy couriers in Italy and the UK", *Work, Employment and Society*, 34 (1), 35–54.

Wood, A. J., Graham, M., Lehdonvirta, V., & Hjorth, I. (2019), "Good gig, bad gig: Autonomy and algorithmic control in the global gig economy", *Work, Employment and Society*, 33 (1), 56–75. doi: 10.1177/0950017018785616.

Woodcock, J., & Graham, M. (2020), *The Gig Economy: A Critical Introduction*, Cambridge; Medford, MA: Polity.

Index